W9-AVC-089

THE TERRY LECTURES

The New Universe and the Human Future

OTHER VOLUMES IN THE TERRY LECTURES SERIES AVAILABLE FROM YALE UNIVERSITY PRESS

The Courage to Be Paul Tillich

Psychoanalysis and Religion Erich Fromm

Becoming Gordon W. Allport

A Common Faith John Dewey

Education at the Crossroads Jacques Maritain

Psychology and Religion Carl G. Jung

Freud and Philosophy Paul Ricoeur

Freud and the Problem of God Hans Küng

Master Control Genes in Development and Evolution Walter J. Gehring

Belief in God in an Age of Science John Polkinghorne

Israelis and the Jewish Tradition David Hartman

The Empirical Stance Bas C. van Fraassen

One World: The Ethics of Globalization Peter Singer

Exorcism and Enlightenment H. C. Erik Midelfort

Reason, Faith, and Revolution: Reflections on the God Debate Terry Eagleton

Thinking in Circles: An Essay on Ring Composition Mary Douglas

The Religion and Science Debate: Why Does It Continue?
Edited by Harold W. Attridge

Natural Reflections: Human Cognition at the Nexus of Science and Religion
Barbara Herrnstein Smith

Absence of Mind: The Dispelling of Inwardness from the Modern Myth of the Self
Marilynne Robinson

Islam, Science, and the Challenge of History Ahmad Dallal

The New Universe and the Human Future

How a Shared Cosmology Could Transform the World

NANCY ELLEN ABRAMS AND JOEL R. PRIMACK

Yale UNIVERSITY PRESS

New Haven & London

Published with assistance from the foundation established in memory of
Philip Hamilton McMillan of the Class of 1894, Yale College.

Yale University Press books may be purchased in quantity for educational, business,
or promotional use. For information, please e-mail sales.press@yale.edu
(US office) or sales@yaleup.co.uk (UK office).

Designed by James J. Johnson.
Set in Linotype Centennial by Tseng Information Systems, Inc.
Printed in the United States of America.

Library of Congress Cataloging-in-Publication Data

Abrams, Nancy Ellen, 1948–
The new universe and the human future : how a shared cosmology could
transform the world / Nancy Ellen Abrams, Joel R. Primack.
p. cm. — (The Terry lectures series)
Includes bibliographical references and index.

ISBN 978-0-300-16508-1 (hardback)
1. Cosmology—Philosophy. 2. Life—Origin—Philosophy. I. Primack, J. R. (Joel R.) II. Title.
QB981.A343 2011
523.101—dc22
2010044895

A catalogue record for this book is available from the British Library.

This paper meets the requirements of ANSI/NISO Z39.48–1992
(Permanence of Paper).

10 9 8 7 6 5 4 3 2 1

To Samara Bay and her fellow artists,

in whose hands rests the portrayal of

the new universe and the human future

The Dwight Harrington Terry Foundation Lectures on Religion in the Light of Science and Philosophy

The deed of gift declares that "the object of this foundation is not the promotion of scientific investigation and discovery, but rather the assimilation and interpretation of that which has been or shall be hereafter discovered, and its application to human welfare, especially by the building of the truths of science and philosophy into the structure of a broadened and purified religion. The founder believes that such a religion will greatly stimulate intelligent effort for the improvement of human conditions and the advancement of the race in strength and excellence of character. To this end it is desired that a series of lectures be given by men eminent in their respective departments, on ethics, the history of civilization and religion, biblical research, all sciences and branches of knowledge which have an important bearing on the subject, all the great laws of nature, especially of evolution . . . also such interpretations of literature and sociology as are in accord with the spirit of this foundation, to the end that the Christian spirit may be nurtured in the fullest light of the world's knowledge and that mankind may be helped to attain its highest possible welfare and happiness upon this earth." The present work constitutes the latest volume published on this foundation.

Contents

Acknowledgments

A great thank you to Priyamvada Natarajan and the other members of the Terry Lectures committee at Yale University for so graciously hosting us in New Haven in October 2009 for the two weeks of our Terry Lectures, and to Lauralee Fields for making it all run amazingly smoothly. More thanks to our agent, Douglas Carlton Abrams, whose advice and help are invaluable, and to our wonderful and tireless assistant, Nina McCurdy, for her careful work assembling our many figures and artistically creating several of them. Thanks and thanks and ever thanks to our brilliant daughter, Samara Bay, whose editing has been a window not only into the English language but also into the artistry of storytelling and the outlook of her generation. To our manuscript editor, Laura Jones Dooley, much appreciation for her careful work and bubbling enthusiasm. And finally, to our dream editor at Yale University Press, Jean E. Thomson Black, who has understood the deepest goals of this book from the start and fought for us: we know how lucky we are, and we're immensely grateful.

Introduction

There is a gaping hole in modern thinking that may never have existed in human society before. It's so common that scarcely anyone notices it, while global catastrophes of natural and human origin plague our planet and personal crises of existential confusion plague our private lives. The hole is this: we have no meaningful sense of how we and our fellow humans fit into the big picture. Are we the handiwork of a loving God who planned the universe? Are we insignificant motes marooned on a lonely rock in endless space? In every culture known to anthropology, people could have answered questions like these with confidence, even though their answers would probably seem quaint or absurd to us now. They knew what their cosmos was like because they lived in a world where everyone around them shared it. We don't.

Despite all we've gained in this scientific age, we've lost something important. Even in a roomful of neighbors, it's highly unlikely that everyone will have the same mental picture of large-scale reality and even more unlikely that any of their pictures is based on real evidence. We're divided on the most fundamental question of any society: what universe are we living in? With no consensus on this question and no way even to think constructively about how we humans might fit into the big picture, we have no big picture. Without a big picture we are very small people.

Many religious believers are convinced that the earth was created as is a few thousand years ago, while many people who respect science believe that the earth is just an average planet of a random star in a universe where no place is special. *Neither is right.* Both groups are operating within mental pictures of the universe that we now know scientifically are wrong. Meanwhile, global problems are escalating—religiously justified brutality, exhaustion of planetary resources, climate chaos, economic disasters, and more. We function day to day in a high-tech, fast-paced world, but modern technology for billions of users is essentially magical. Astronomy appears to have little relevance. People think of astronomical discoveries as inspiration for kids or a great topic for five minutes of clever dinner party banter, but there's no widely understood connection between what's happening in distant space and us, right here. The truth is, however, that there is a profound connection between our lack of a shared cosmology and our increasing global problems. Without a coherent, meaningful context, humans around the world cannot begin to solve global problems together. If we had a transnationally shared, believable picture of the cosmos, including a mythic-quality story of its origins and our origins—a picture recognized as equally true for everyone on this planet—we humans would see our problems in an entirely new light, and we would almost certainly solve them. Getting from here to there is what this book is about.

By incredibly fortunate coincidence, there's a scientific revolution occurring today in the branch of astrophysics called "cosmology," and it's revealing our true cosmic context. The meaning of this earth-shaking discovery could transform our minds and thus our world.

Scientific cosmology is the study of the universe *as a whole*—its origin, nature, and evolution. This field has made remarkable progress since the end of the twentieth century in figuring out how the universe works, even though the universe is almost entirely invisible. Everything we can see with all our instruments is less than 1 percent of what's actually out there. Most of the matter in the universe is invisible "dark matter," as scientists have termed it. This and other fundamental discoveries may make it possible to figure out how the universe operates on *all* size and time scales, including our own. From this new understanding we are coming to see how we humans fit into the scientific big picture, what our significance is within that context, and why our current actions here on Earth matter far beyond most people's imagination.

Religious origin stories like the ones in Genesis have never been factually correct about the universe (for example, it was not made in six days), but they served as a cultural cosmology. In lieu of scientific accuracy they offered guidance about how to live with a sense of belonging and how to draw strength from feeling part of a larger, shared presence that could give life's more mundane moments meaning. Modern scientific cosmology, in contrast, says much about dark matter and the workings of the universe but nothing about how human beings should live or feel. It aims to provide scientific accuracy but not meaning or guidance in life.

This split between science and human meaning is not a reflection of reality, however, but the result of a historical choice made four centuries ago. When Galileo was arrested and convicted by the Catholic Church for teaching that the earth moves, it was a sobering event for scientists all over Europe because Galileo was the greatest scientist of his time and the first scientific celebrity. As a result, leading scientists such as René Descartes adopted—for their own protection—a policy of noninterference with religion: they would make no claims to authority over anything but the material world; they would defer to religion in all questions of meaning, value, and spirit. The church fathers, on the other side, needed to protect themselves from endless battles over future scientific discoveries and the embarrassment of seeing their religious doctrines subverted. They accepted this "Cartesian Bargain," and the arrangement has been helpful in allowing science to flourish, especially in past centuries. But given the enormous and pressing global issues that confront us, the modern world can no longer afford to maintain this historical fiction and see fact and meaning as automatically separate. We cannot afford to have an accurate scientific picture on the one hand while on the other being guided in our feelings, philosophies, and views of the future by ancient fantasies that stand in for fact but have long since been disproved—because that's in fact what we've been doing. The human race needs a coherent, believable picture of the universe that applies to all of us and gives our lives and our species a meaningful place in that universe. It's time to reconnect the two different understandings of the word *cosmology*—the scientific and the mythic—into one: a science-based appreciation of our place in a meaningful universe.

In this book we present the new scientific picture of the universe as visually as possible, but we then venture beyond science and suggest what accepting this new picture might mean for our lives. Our goal is to show how our society might begin to conquer seemingly intractable global problems by filling the gaping hole in our thinking, applying these new ideas, and eventually becoming a new global society with a common origin story.

By helping us come to terms with our place in a dynamic, evolving universe where time is measured in both billions of years and nanoseconds and size is measured both across great galaxy clusters and across the nucleus of an atom, the new cosmology gives us the concepts we need in order to begin thinking in, and acting for, the very long term. It lets us appreciate our significance to the universe as a whole. One of the most frightening problems facing our world today is the large number of people who are using sophisticated, high-tech weapons to violently impose their regional rivalries and narrow religious notions on the whole world—in short, people who are acting globally but thinking locally. This is backwards: we have to think on a larger scale than the one we're acting on, if those actions are to be wise. *To act wisely on the global scale, we need to think cosmically.* And for the first time this is possible.

Earth is incredibly special, more so than anyone imagined before recent discoveries of hundreds of other planets orbiting nearby stars. And our era is an incredibly special moment even on a timescale of billions of years: we are the first species that has evolved with the capability to destroy our planet. Will we do so? Or will we successfully negotiate over the next two generations a transition from exponential growth in environmentally harmful activities to a sustainable relationship to this remarkable planet, the only hospitable place for creatures like us in the explored universe? The answer could affect not only humanity but the entire future of intelligence in the ultimately visible universe.

A generation ago, the word *cosmic* was only hazily suggestive. Since no one knew what the cosmos actually was, *cosmic* had no precise meaning. When used to modify words like *consciousness* or *perspective* or *identity,* the word *cosmic* seemed flaky if not ridiculous. But today we are beginning to know what the cosmos is from the horizon of the universe to a single elementary particle, so the word *cosmic* can now be understood not in its old, vaguely disreputable sense but as referring *specifically* to the new scientific picture of the universe. Cosmic consciousness is consciousness arising in this universe—whether human or alien—encompassing the new universe picture, and accepting and working from its principles. Cosmic identity is our own identity, based on the specific and fundamental ways that we fit into this new picture. In other words, the word *cosmic* is now legitimate in a way it has not been since the dawn of modern science.

But this book is not about science per se. It's about us and what we as a species need to do, now that we understand for the first time where we are in time and space. It explains the minimum amount of cosmology needed to get across to any interested person, no special background required, the new meaning of "where we are." If your scientific curiosity is still unsatisfied, please have a look at the Frequently Asked Questions section at the back of this book or read our earlier book, *The View from the Center of the Universe: Discovering Our Ex-*

traordinary Place in the Cosmos. The real focus of this book is on the invitation, and in fact the imperative, to free our society from obsolete, dangerous misconceptions of physical reality, open our minds to the new universe, and begin to teach and cultivate the exciting connections between our universe and both our internal sense of power and our external, political outlooks. In short, this is an invitation to create a "cosmic society."

Many of the pictures in this book are frames taken from videos that are based on supercomputer simulations of key aspects of the evolving universe. You can watch these videos on our website http://new-universe.org whenever you see the symbol ▣. There are also additional videos on the website, listed by the chapter of this book to which they are relevant.

The long-term success of our species may very well depend upon our becoming a cosmic society, capable of thinking on the grand scale while acting on the small. A cosmic society is not about zipping around the Galaxy visiting aliens—it's about expanding our thinking and transforming our actions right here on Planet Earth. It's radical but simple, and for the first time in human existence it's within our reach.

Fig. 1. Great telescopes on and orbiting Earth

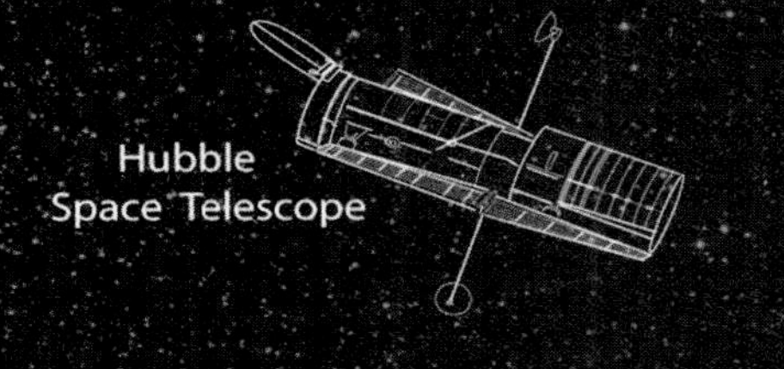

Chapter 1
The New Universe

Scientists used to joke that cosmology was the field where the ratio of theory to data was nearly infinite. Lots of theories, almost no data. But over the past two decades that ratio has flipped: it's now nearly zero. A huge and ever-increasing amount of data has ruled out all theories but one, and that one theory not only fits all available data, it's been predicting the data. This theory is the foundation of a new picture

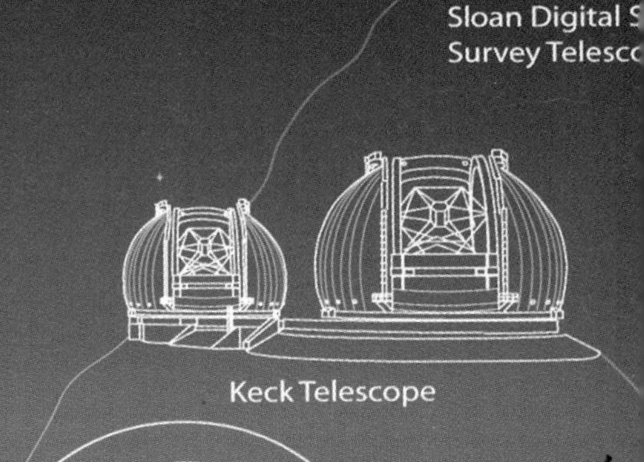

Fig. 2. A pillar of star birth: The Carina Nebula in visible light

Fig. 3. A pillar of star birth: The Carina Nebula in infrared light

of the cosmos, our new universe. Its technical name is Lambda CDM, but it's more simply called the Double Dark theory.

In the new understanding, everything visible with the greatest telescopes on and orbiting Earth is only half of 1 percent of what the universe actually contains.

The stars, the glowing gas clouds (figs. 2, 3), the planets (fig. 4), and the hundred billion galaxies out to the cosmic horizon (fig. 5) are all in the visible half a percent. What's the rest? The universe *as a whole* is controlled by two invisible things whose tango with each other since the Big Bang has created the visible galaxies—the only homes for the evolution of planetary systems and life. The two invisible dancers are *dark matter* and *dark energy,* the two "darks" of the Double Dark

Fig. 4. Saturn with Earth in the background

Fig. 5. The Hubble Ultra Deep Field in infrared light

theory. Despite their overwhelming importance to the universe as a whole, dark matter and dark energy were unknown and even unimagined until the twentieth century. Between dark matter collapsing the galaxies into being and dark energy accelerating them apart from one another, our evolving universe has turned out to be far more dynamic than the old picture of endless space scattered with stars.

What is emerging from modern cosmology is humanity's first scientifically accurate story of the nature and origin of the universe.[1] Building on the great scientific achievements of the nineteenth and early twentieth centuries—particularly evolution, relativity, and quantum mechanics—the end of the twentieth century and the beginning of the twenty-first have been a golden age of astronomy. In a very real sense, we have discovered the universe.

The new scientific picture differs from all earlier creation stories not only because it's based on evidence but also because it's the first ever created by collaboration among people from different religions, races, and cultures all around the world, each of whose contributions is subject to the same standards of verifiability. The new universe picture excludes no one and sees all humans as equal. It belongs to all of us, not only because we're all part of it but also because around the world the work to discover it has been largely funded by the public. The fruit of this transnational collaboration could become a unifying, believable picture of the larger reality in which Earth, our lives, and the ideas of all our religions are embedded.

In the entire recorded history of Western civilization, there have been only three fundamentally different physical pictures of the uni-

Fig. 6. The ancient Egyptian cosmos

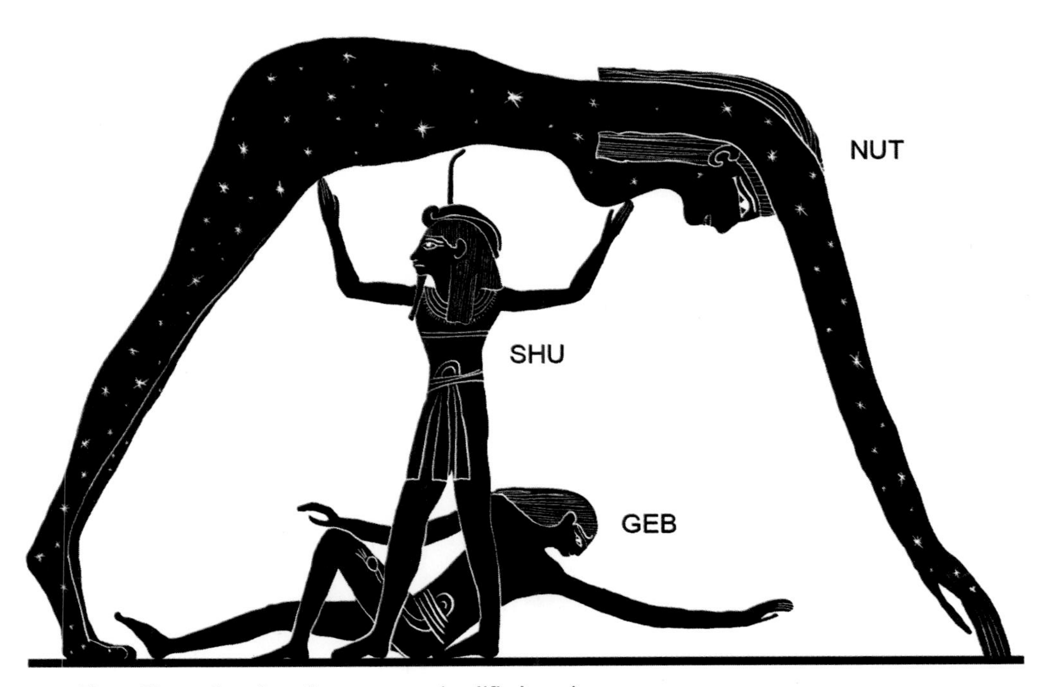

Fig. 7. The ancient Egyptian cosmos, simplified version

verse, although within each of these there have been countless variations. In the earliest picture the earth was flat. Let's look, for example, at a representation of the cosmos of ancient Egypt (fig. 6).

Paring this down to a simplified version, we see that the flat earth is a god named Geb (fig. 7). Heaven is his sister and lover, the goddess Nut (pronounced Noot), whose body contains the stars. Nut and Geb, heaven and earth, were born locked together in an embrace and were pried apart by their father, Shu, the god of air or space. They longed to come together again, but Shu held them apart, thus maintaining the space between heaven and earth. Egyptians believed that he would do this only as long as they continued to perform their rituals properly, day after day. If they abandoned their religion, heaven and earth would come together again and that would be the end of the creation. So in practicing their religion, ancient Egyptians felt that they were literally upholding the cosmos, and this gave them a sense that they truly mattered.

In the Egyptians' big picture the land was embedded within an awe-inspiring but mostly invisible cosmos filled with gods. The waters of the Nile flowed in from the spiritual world onto farms. The Egyptian people had descended from the gods. People often painted Nut on the inside of coffin lids, so that the deceased would lie under the protective presence of Nut and be welcomed back to her.

The stories of the gods—not only Geb, Nut, and Shu but many, many others—explained to the ancient Egyptians why things worked as they did. Their cosmology was more complicated than this and came in many local versions, but all versions had this in common: they provided a rich and satisfying explanation of life, nature, cosmos, and the gods, even though the nature and cosmos parts were far from accurate by modern standards.

Much later in ancient Israel and Judah, the flat earth cosmos still reigned unquestioned (fig. 8). The three-part structure of heaven, earth, and space in between was the same for the Hebrews as for the Egyptians; however, the parts were no longer gods but inanimate earth, air, and firmament, since for the Hebrews there could be only one God.

Then came Western history's first great cosmological shift. The

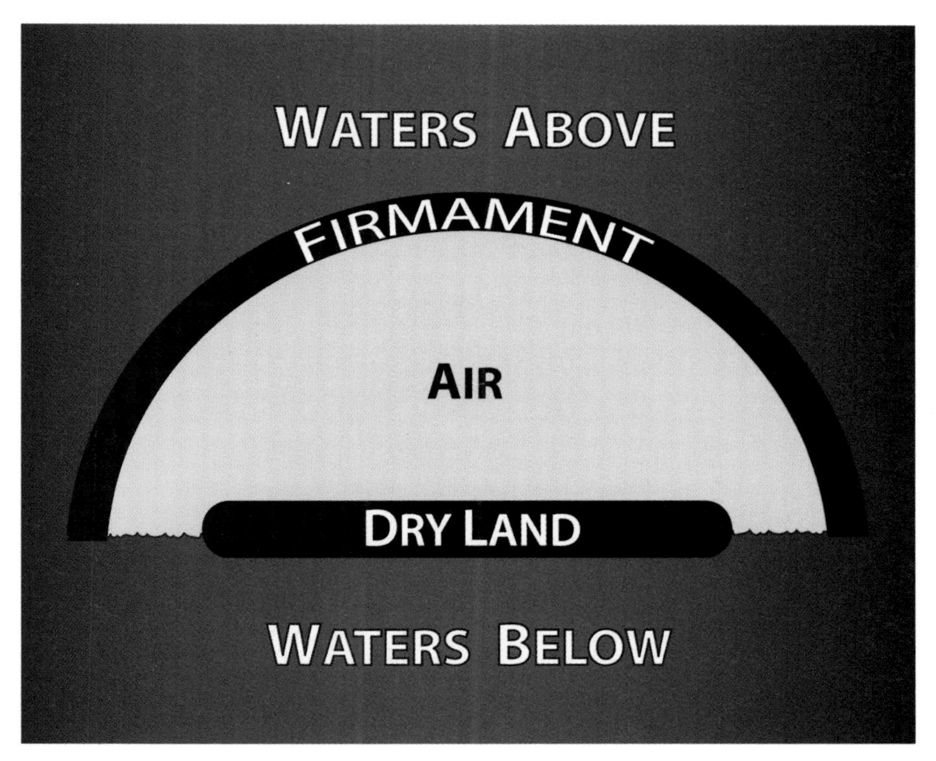

Fig. 8. The ancient Hebrew cosmos

ancient Greek imagination burst out of the two-dimensional flatland earth of the Middle East to a three-dimensional universe. The Greeks realized that the earth is not a pancake sitting on water but a sphere surrounded by space. Over the next thousand years, this idea slowly spread until by the Middle Ages, educated people from the Middle East and North Africa across Europe to Scandinavia had all come to believe that the earth was a sphere at the center of a spherical universe (fig. 9).

Everything in the sky was now understood to revolve around the earth. Nested crystal spheres carried the moving planets, the moon, and the sun, while the outermost sphere carried the fixed stars. The whole universe rotated around the earth once a day, and outside the sphere of the fixed stars was Heaven. In the medieval Christian view, God had placed the spheres exactly where they belonged in a "great chain of being," and the place of every creature and institution on the earth was part of the downward continuation of the cosmic hierarchy of the spheres. It was considered blasphemous to question the hierarchy or your place in it, because it was the place God had chosen for you. As in Egypt, this medieval cosmology both explained and enforced the social rigidity of the time. The medieval picture, like the Egyptian, provided the common folk a satisfying explanation of their existence. The ordinary world was seen as surrounded by a spiritual cosmos, and the doings and expectations of the spiritual beings who lived there gave meaning to daily life here. This was the cosmic picture so beautifully described by Shakespeare in *Cymbeline* (act 5, scene 5), written, ironically, at the same time that Galileo was disproving it:

> The benediction of these covering heavens
> Fall on their heads like dew! for they are worthy
> To inlay heaven with stars.

Fig. 9. The medieval cosmos

The second great cosmological shift began in 1543 when a churchman named Nicolaus Copernicus proposed that if the sun rather than the earth were seen as central, it would be easier to understand the motions of the planets. Over the next century Galileo, Johannes Kepler, and Isaac Newton completed the Copernican revolution by showing observational evidence that the earth could not be the center of the universe and working out the physics of a sun-centered universe. This new cosmic picture caused tremendous controversy, but it took hold because of its predictive and explanatory power. The medieval picture was replaced by what we now refer to as the Newtonian picture—the cosmos of the Enlightenment.

In the Newtonian picture, Earth is not the center of the universe. There is no center. Instead Earth is a planet, and it moves like the other planets, orbiting our star, the sun. No place in the universe is special or central, least of all ours. M. C. Escher's intriguing drawing *Cubic Space Division,* although it shows no celestial objects, clearly portrays the fundamental Newtonian idea of space as an endless grid, in which no place is different from any other (fig. 10). For the first time in the history of cosmologies, there was no longer a place where the physical became the spiritual. It was physical all the way out, possibly to infinity.

The Newtonian picture has only become more dominant since the seventeenth century. Religious people in the intervening centuries have often felt faced with a choice: either (1) to deny the validity of science and isolate themselves from much of the human endeavor, or (2) to adopt a dualistic view in which they accept science but believe that there are two kinds of reality, the spiritual and the physical, with "spiritual reality" unconstrained by the laws of physics. In such a split mind-

Fig. 10. The Newtonian cosmos, as represented by M. C. Escher's *Cubic Space Division*

set it's possible to believe in science and still believe in almost anything else. Fortunately, (1) and (2) are no longer the only alternatives.

We are living at the third great cosmological shift, a scientific revolution that is likely to be as profound in its consequences as the Copernican revolution. Today's revolution began early in the twentieth century, when most astronomers believed that *the Milky Way was the*

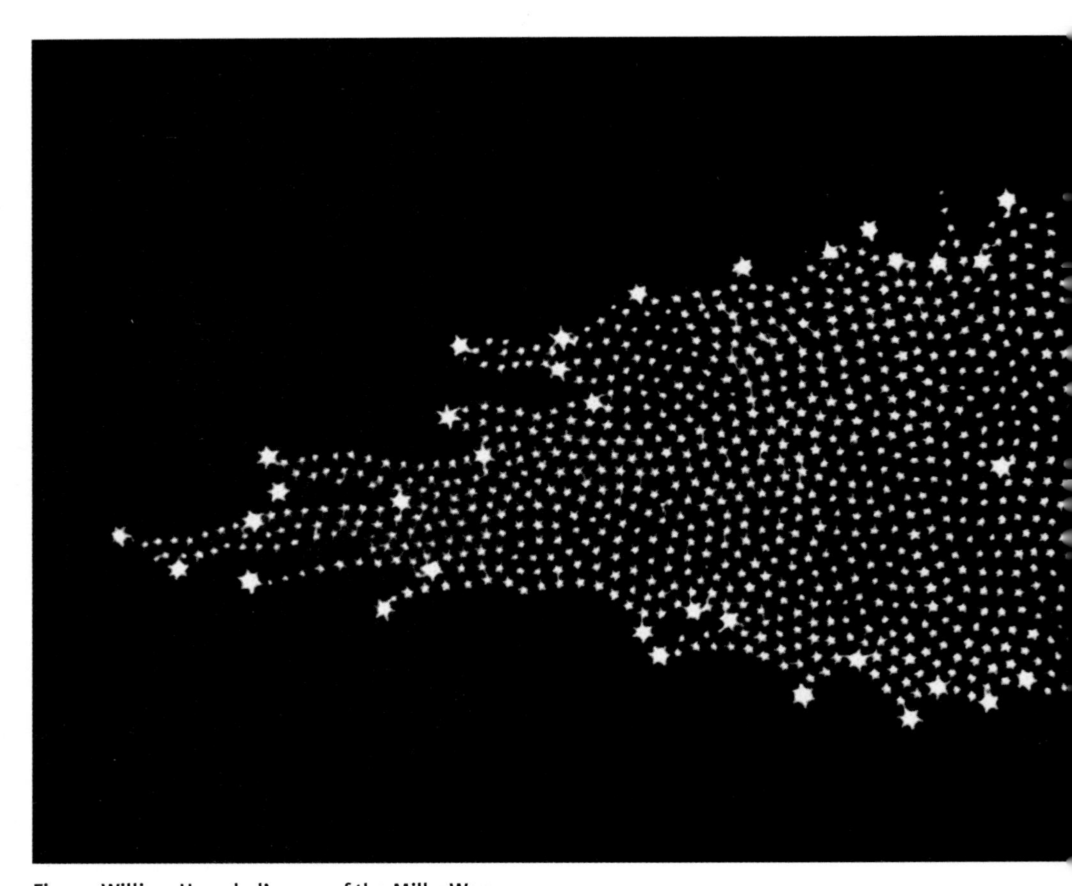

Fig. 11. William Herschel's map of the Milky Way

universe. In the eighteenth century, for example, British astronomer William Herschel mapped the Milky Way with the sun located incorrectly near the middle (fig. 11). In those days all blurry objects that astronomers observed were called *nebulae,* meaning clouds—that is, they were thought to be clouds of gas inside the Milky Way. But American astronomer Edwin Hubble discovered in 1924 that some of the nebulae are actually galaxies far outside the Milky Way. Suddenly our Galaxy

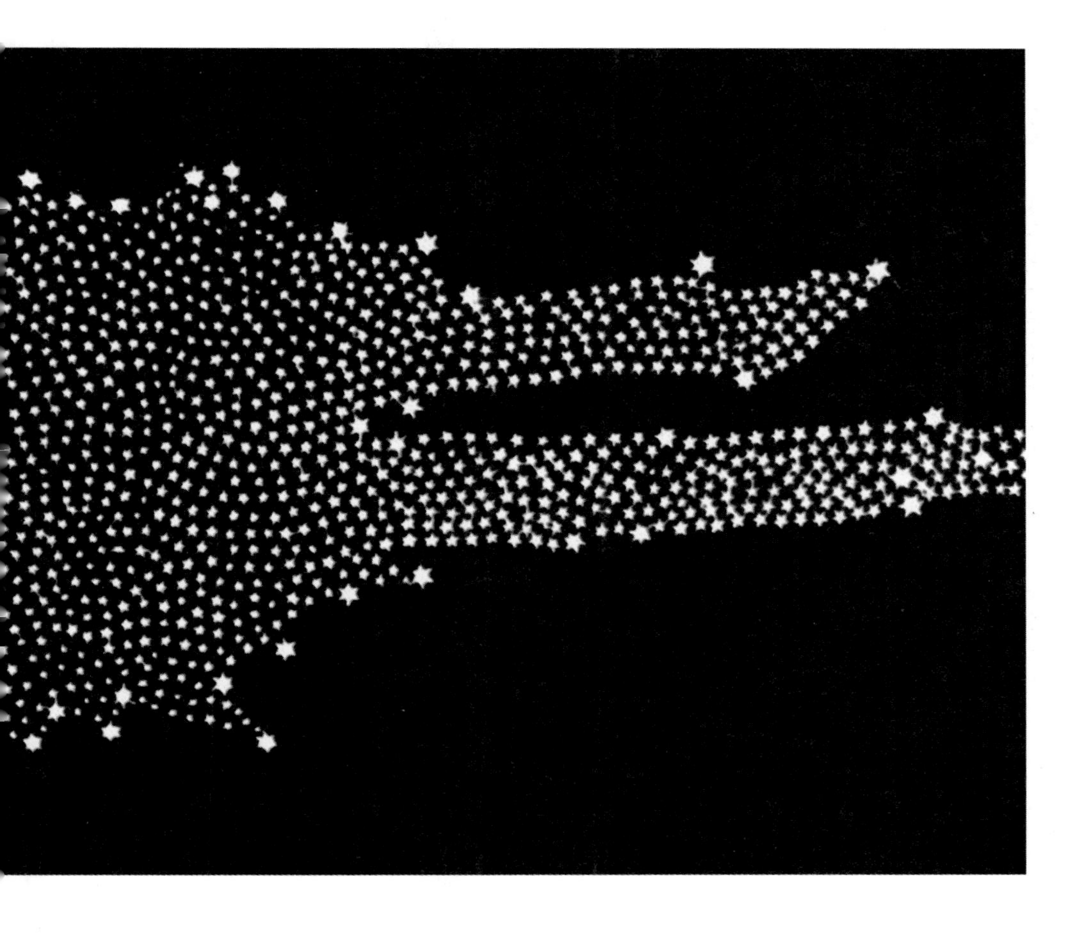

became just one among billions, and the universe became incalculably vast. Then in 1929 Hubble made an even more stunning discovery: the distant galaxies are all moving away from us, and the farther away they are, the faster they're moving. Hubble had discovered the expansion of the universe. Reasoning backwards, astronomers realized that if the universe is expanding now, there had to have been a time when everything was much closer together. That was the Big Bang, but there was no direct evidence for it yet.

By the mid-twentieth century there was lively debate in astronomy between the Big Bang theory and the Steady State theory. Both theories accepted the evidence that the universe is expanding, but the Steady State proponents argued that there was no Big Bang; the universe had no beginning and is basically unchanging, with matter being spontaneously created to form new galaxies as the old galaxies expand apart. The Steady State theory, however, was seriously undermined by the discovery in 1965 that faint heat radiation from the Big Bang fills the universe. Steady State was then dealt a deathblow by the discovery that very distant galaxies (in time as well as space) don't look like nearby ones; this showed that the universe has been evolving. The Big Bang became generally accepted as far as it went, but it could not explain why galaxies formed from a universal explosion. It was decades before the Double Dark theory offered an explanation.

By about 1980 most astronomers had become convinced that most of the mass holding galaxies and clusters of galaxies together is invisible, but they did not understand how that could work. In 1993 came the first solid evidence that the cold dark matter part of the Double Dark theory might be right. Then in 1998 astronomers discovered that the expansion of the universe is actually accelerating; this was the first real confirmation of dark energy. Since then all the rapidly accumulating astronomical evidence has supported the Double Dark picture.

This revolution is presenting us with an opportunity so rare that it has arisen only twice before—the opportunity to reenvision reality itself at the dawn of a new picture of the universe. Now the big question is, what will our culture do with this knowledge? We will come to this. First, however, we need to understand the new picture, and anyone with an open mind can.

Our cosmic address is best described by starting from home and moving outward (fig. 12). Of the eight planets in our solar system, Earth is the third from the sun—not too hot and not too cold. Light crosses the solar system in just a few hours, so the solar system is a few "light hours" across. But it takes light about a hundred thousand years to cross our Galaxy, the Milky Way. (The speed of light in space is 300,000 kilometers per second, or 186,000 miles per second, and it's fixed. A light-year is the distance light travels in a year.) Our entire solar system is a tiny dot about halfway between the visible edge of the Milky Way and the central bulge of stars.

The Milky Way is part of what astronomers call the "Local Group" of galaxies; it's a group held together by gravity, consisting of two big galaxies—ours and the Andromeda galaxy—plus about fifty smaller satellite galaxies (with more little ones now being discovered by increasingly powerful telescopes). Our Local Group is just a dot on the scale of the "Local Supercluster" of galaxies—about a thousand bright galaxies (and many thousands of additional faint ones) spread out in a thick sheet about a hundred million light-years across and still expanding apart. Embedded in the Local Supercluster, about sixty million

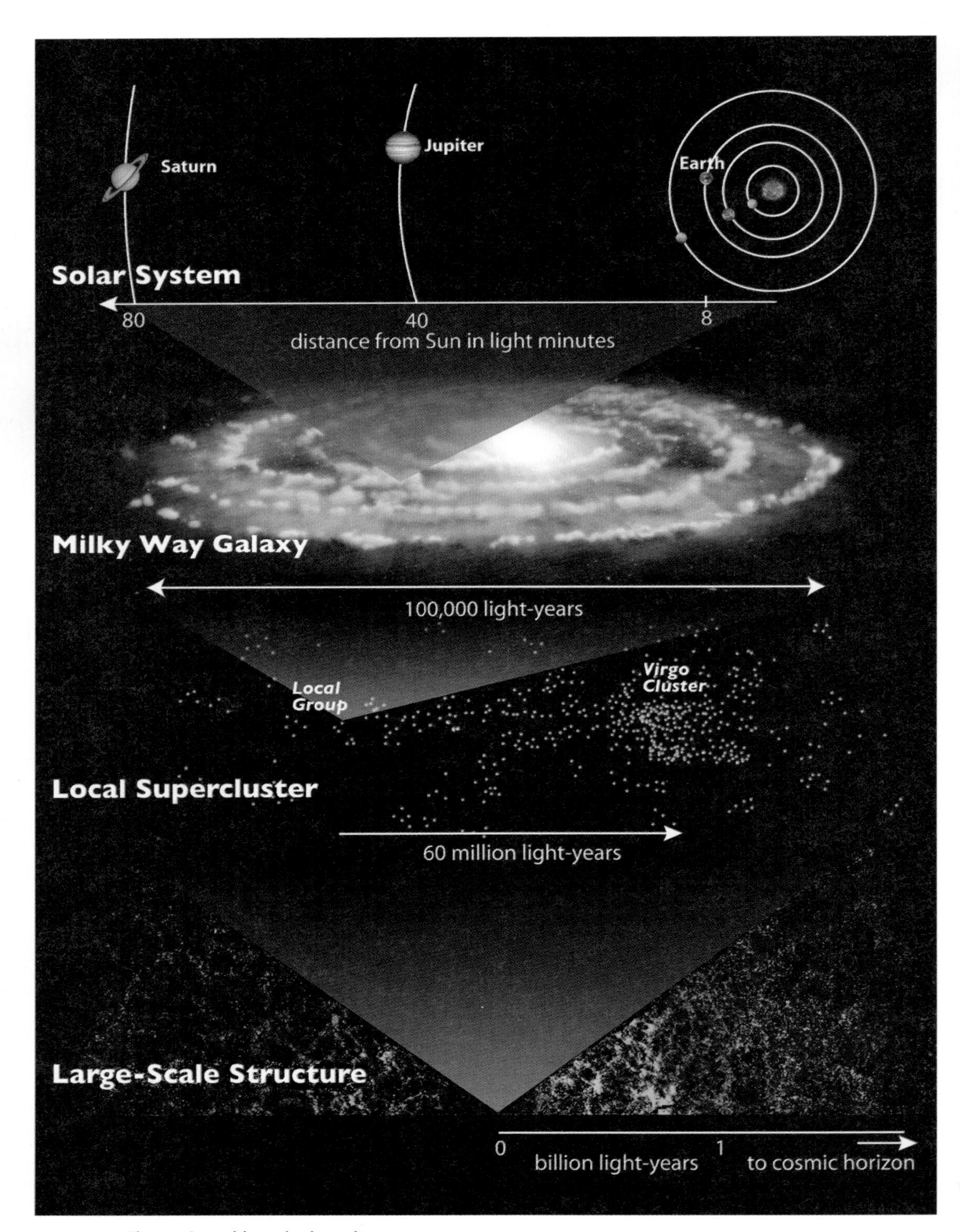

Fig. 12. Our address in the universe

light-years from us, is the "Virgo Cluster" of galaxies. It is named that because we see it in the direction of—but through—the constellation Virgo. All constellations are made of foreground stars near the sun inside our own Galaxy.

In the sequence of pictures below, we take a virtual voyage starting from Earth, flying past actually mapped local stars, out of the disk of the Milky Way, and then partway across our Local Supercluster to the Virgo Cluster, a distance of sixty million light-years. You can watch this ▣ Voyage to the Virgo Cluster as a video on our website, http://new-universe.org.

Beside the constellation Orion as seen from Earth, the disk of the Milky Way arches across the sky to the left (fig. 13). As we fly toward the sword hanging below Orion's belt, the sword comes apart, because all the stars making it up are at different distances. Getting closer, we see that the center of the sword is not a star but the Orion Nebula, a gas cloud illuminated by the young stars forming there (fig. 14).

But this is all inside the disk of the Milky Way, where dust clouds block part of the light. If we rise above and outside the disk, we can see the full panorama of our Galaxy with its hundreds of billions of stars (fig. 15). The Milky Way has two nearby satellite galaxies, called the Magellanic Clouds, visible just to its left. All the little patches of light in the background are not stars but galaxies, many of which are as bright as the Milky Way. All these galaxies are in our Local Supercluster.

Clusters of galaxies like the Virgo Cluster are found at the intersections of chains or filaments of galaxies. The Virgo Cluster is shown here with a long chain of galaxies stretching to the right (fig. 16). Most of the galaxies in the chain are disk galaxies (fig. 17) like the Milky Way, but in the Virgo Cluster there are also elliptical galaxies—big balls of

Fig. 13. The Orion Constellation as seen from Earth

stars without a disk. At the center of the Virgo Cluster is the giant elliptical galaxy M87 (fig. 18). This galaxy has a black hole at its center that weighs more than three billion times the mass of our sun. Some of the material that fell toward the black hole is being shot out as a jet.

Yet even sixty million light-years is a relatively short trip compared to ones the Double Dark theory is now allowing astronomers to simulate. These huge numbers often make people feel insignificant in comparison. This feeling is so common in our culture, even among children, that here it is in both Peanuts and Calvin and Hobbes (figs. 19, 20). In both cartoons the feeling of cosmic insignificance is uncomfort-

Fig. 14. The Orion Nebula

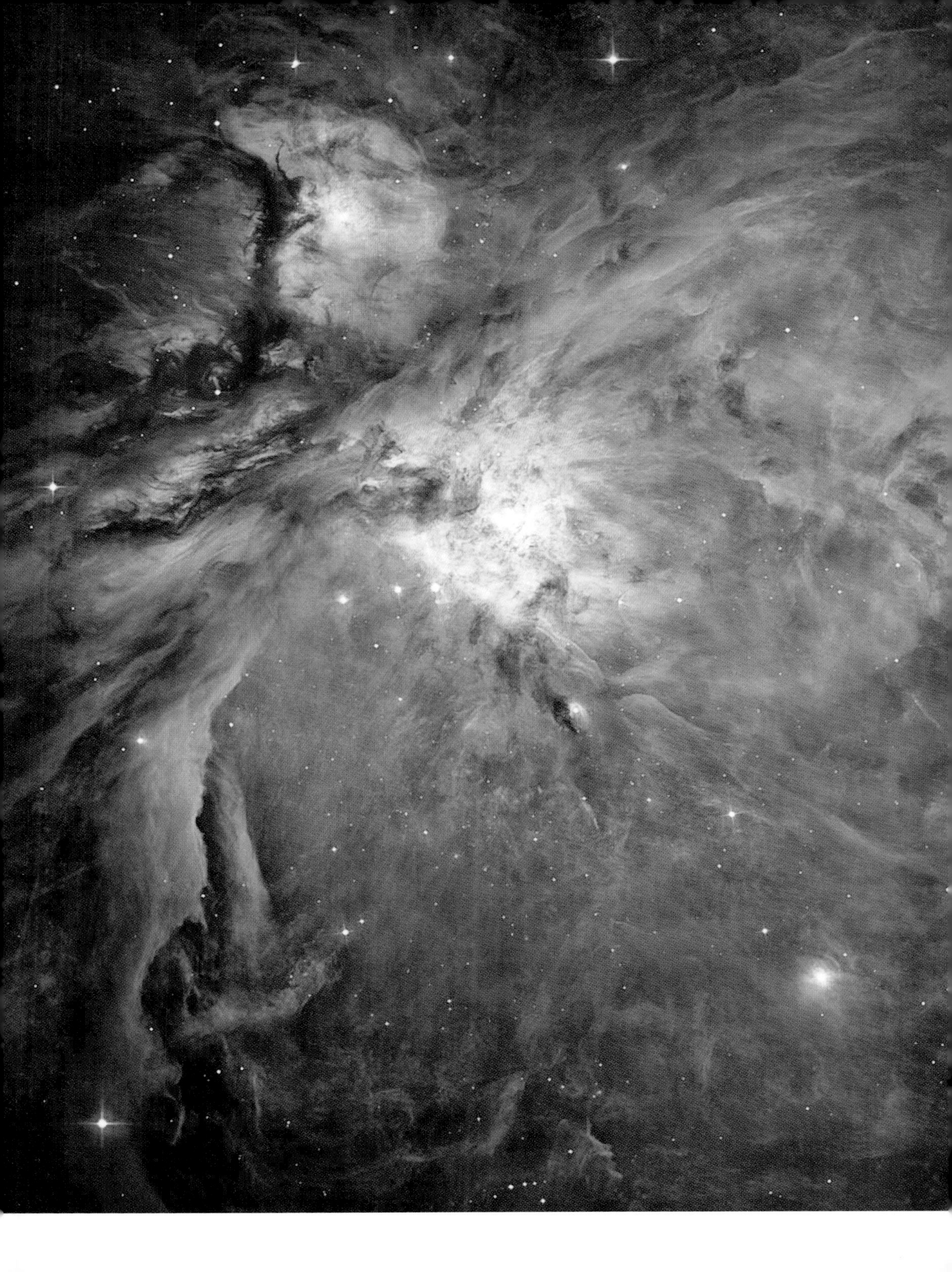

Fig. 15. The Milky Way galaxy with large and small Magellanic clouds

Fig. 16. The Virgo Cluster and a chain of galaxies

Fig. 17. The Whirlpool Galaxy (M51)

Fig. 18. Galaxy M87 at the center of the Virgo Cluster

Fig. 19. Peanuts: "You are of no significance"

Fig. 20. Calvin and Hobbes: "What a clear night!"

able to think about, so the characters run from it. This feeling derives from the Newtonian assumption that in an incomprehensibly vast, cold universe we are, to quote the famous biologist Stephen Jay Gould, a "fortuitous cosmic afterthought."[2] But we know now that this is not the case. The new picture is revealing a universe in which we intelligent beings have a central or special place in several different senses of the phrase. We have at last a way to visualize the whole, so that from this new perspective we can confidently see our full selves.

Chapter 2
Size Is Destiny

Picture a young girl sitting under a tree on a planet (fig. 21). The girl (as well as the tree) is a community of hundreds of billions of living cells, dividing and keeping life running. And every cell is an entire world within itself, yet even its tiniest parts are made of millions of atoms. Meanwhile there are hundreds of billions of planets in our own Galaxy, and hundreds of billions of other galaxies. Sizes are like these doorways within doorways: when you pass through one, everything changes. But there's a limit.

Pure numbers go on infinitely, but sizes of actual things don't. The universe began a certain amount of time ago and has been expanding at a certain rate; therefore it's gotten to be a certain size, and that's the largest size we can say anything definitive about.

Ok, you might think, but in the other direction can't sizes get infinitely smaller? Can't things be divided in half again and again, at least theoretically? Once again, pure numbers can be divided in half for-

Fig. 21. Sizes are doorways within doorways

ever, but physical reality is different. The interplay of General Relativity and Quantum Mechanics establishes a smallest size, called the "Planck length," and rules out the existence of anything smaller. By understanding how sizes operate, we gain much more than an abstract understanding of the universe as a whole: we gain insight into ourselves.

Most people are still thinking about global politics or economics with understandings, moral judgments, and above all a sense of time that are appropriate only to much smaller size scales like a family or a community, a year or a lifetime. They have no way of comprehending the hundreds or even thousands of human generations that the things now being done to our planet may last. For example, even if everyone around the world completely stops putting greenhouse gases into the atmosphere tomorrow, the oceans will keep rising for at least a thousand years because of climate change that's already occurred. Radioactive waste could be dangerous for a hundred thousand years. Extinction is forever. The *science* of cosmology aims only to understand the underlying principles and history of the universe, not to change people's lives or challenge their beliefs. But once we understand these principles, they necessarily change our thinking.[1] We realize that we are governed by them, and so is everything. That's what "universal" truly means.

Our modern universe can be represented as a continuity of vastly different size scales along the body of a serpent (fig. 22). A serpent swallowing its tail is a very ancient symbol of the cosmos used by countless cultures, and here we follow in this venerable tradition. The tip of the tail represents the smallest size, the Planck length, and the head of the serpent represents the largest, the size of the visible universe.[2]

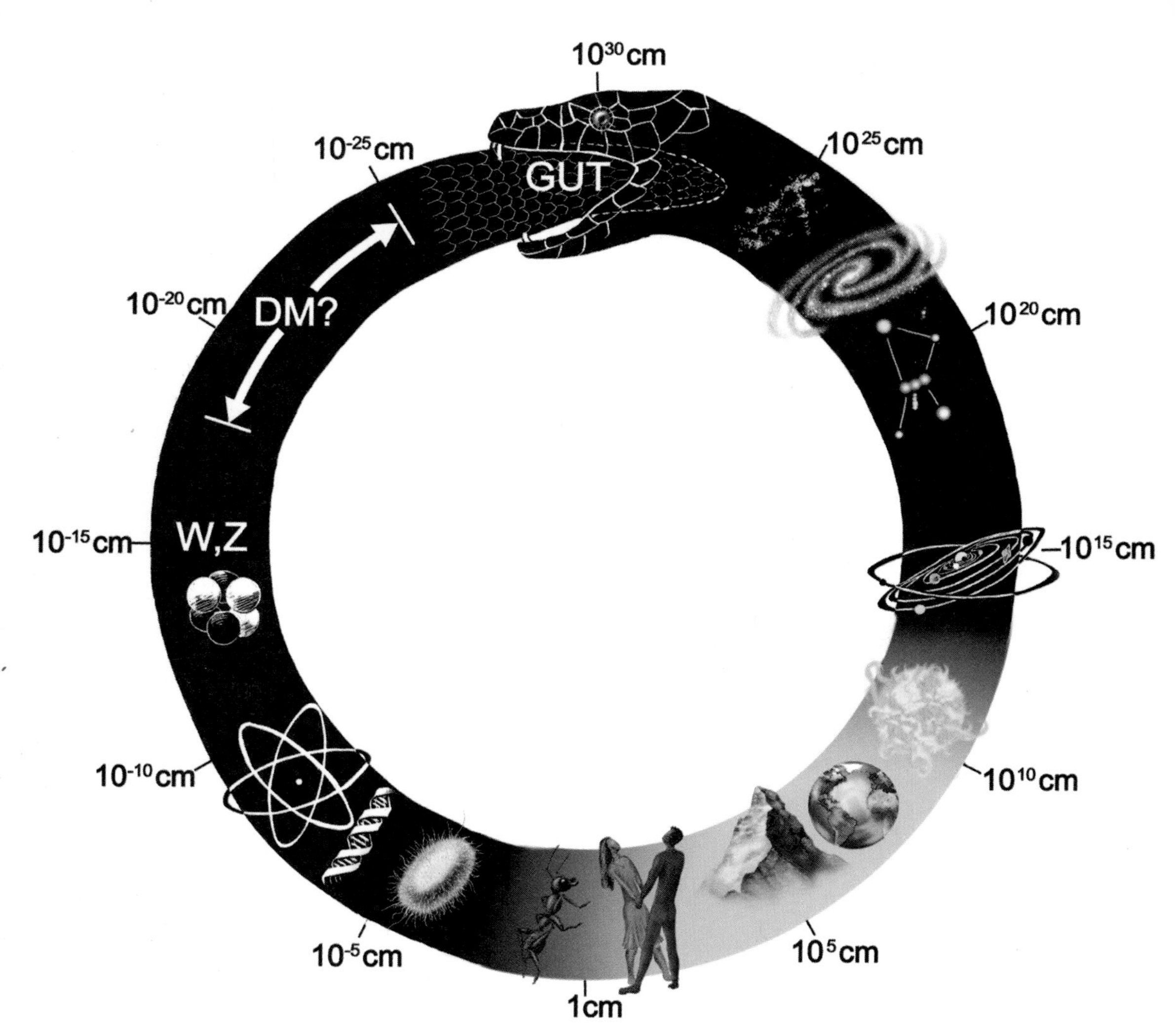

Fig. 22. The Cosmic Uroboros

Uroboros is the ancient Greek word for tail-swallowing, so we call our symbol the Cosmic Uroboros.

The numbers representing sizes change not linearly but by powers of ten. About sixty orders of magnitude—sixty powers of ten—separate the smallest size from the largest. Each tick marks a size a hundred thousand times (10^5) bigger than the tick before. The size scales of the universe can only be shown on a logarithmic scale, since that gives them all equal importance, as indeed they have. Traveling clockwise around the serpent from head to tail, the icons represent ▣:

a supercluster of galaxies (10^{25} cm),
a single galaxy,
the distance between Earth and the Great Nebula in Orion,
the size of the solar system,
the sun,
the earth,
a mountain,
humans,
an ant,
a single-celled creature such as the *E. coli* bacterium,
a strand of DNA,
an atom,
an atomic nucleus,
the scale of the weak interactions (carried by the "W" and "Z" particles),
and, approaching the tail, the extremely small size scales on which physicists hope to find dark matter (DM) particles,
and, on even smaller scales, a possible Grand Unified Theory (GUT) that connects the head to the tail.

Everything in the universe *has* to be about the size it is. Size is destiny because the size of an object or event determines what laws of physics control it. The success of modern physics is based on the assumption that all the laws of physics are always true. However, on different size scales, different laws take control while others become nearly irrelevant. This is why scale models never work like the real thing.

The sizes represented on the Cosmic Uroboros are the only possibilities in our universe. On the largest scales (from about midnight to five if the Uroboros were a clock face) gravity controls. In the middle size range from about five to eight, gravity still matters, but continually less; far more important here is electromagnetism, which is the basis of chemistry and holds material together. As things, or animals, get smaller, gravity's importance diminishes until you can flick an insect off a table and when it lands on the ground it's completely unhurt. At nuclear scales (around nine on the clock face) the strong forces overpower electromagnetism; this is why an atomic nucleus sticks together even though it's made of many positively charged protons that are electromagnetically repelling one another. Gravity at these tiny sizes has long since become irrelevant. But then, approaching the tip of the tail near the Planck length, gravity again becomes the most powerful force. The swallowing of the tail represents the hope physicists have that gravity links the largest and smallest sizes and unifies the universe.[3]

Galileo gave the first clear explanation of scaling. He figured out that no animal could be three times its normal height and be the same shape, because if its height increased by a factor of three, then the strength of its bones (which is determined by their cross-sectional *area,* or thickness) would increase $3 \times 3 = 9$ times; but its weight (determined by the bone's mass) would, like *volume,* increase $3 \times 3 \times 3 = 27$ times. So its weight would crush its bones. Longer bones must be proportionately thicker. This is why an elephant doesn't look like a large gazelle—and why there can't be monstrous insects the size of cars and living children can't be shrunk to the size of mice, despite science fiction movies to the contrary.

Hollywood doesn't seem to care, so the moviemakers actually miss some great opportunities. King Kong in the 2005 movie looks like a normal lowland gorilla in front of a miniature background. But reality is stranger than fiction. Showing King Kong bulked up, as he would have had to look if he were actually as tall as he was supposed to be, would have been far more terrifying. And his fall from the Empire State Building would in reality not merely have left his body lying in the street: it would have splashed pink mush all over Manhattan.

We humans are almost exactly at the center of all possible sizes—at the bottom of the Cosmic Uroboros, halfway between the smallest and the largest. That is our place whether the units used are centimeters, nanometers, or light-years, so this is not merely an anthropocentric bias. If we used different units, the only change would be the numbers in the exponents all the way around the serpent; we would still be central.

And we couldn't be anywhere else. If we were much smaller, we wouldn't have enough atoms to be complex. If we were much larger, the speed of thought and other internal communications (which are limited by the speed of light) would be too slow. Only near the center of all possible sizes can consciousness as complex as ours arise, and this tells us something important about intelligent life anywhere in the universe: if it exists at all, it will have to be approximately our size, somewhere between a redwood tree and a puppy, which is a very narrow range of possibilities.[4]

When we humans look around from our vantage point at the center of the Cosmic Uroboros, the size scales surrounding us make up our conscious world, the section shown in light blue in figure 22. It includes everything from the littlest creatures visible with the naked eye, smaller than ants, up to the sun. This section is what most people think of as reality—but it's not all of reality; it's only about a quarter of it. Yet this section is special: it's the part of the universe for which we human beings have awareness and intuition. This is our mental homeland in the universe. We can call it Midgard.

In the Old Norse cosmology, Midgard was the human world; it was an island representing stability and civilized society in the middle of the world-sea, the Norse universe. In one direction across the world-sea lay the land of the giants, and in the other, the land of the gods. This is an excellent description—metaphorically, of course—of Midgard on the Cosmic Uroboros. Our Midgard is the island of size scales that are familiar and comprehensible to human beings. But beyond the shores of Midgard moving counterclockwise, outward into the expanding universe, lies the land of incomprehensibly giant beings, like black holes millions of times the mass of the sun and galaxies made of hundreds of billions of stars. Clockwise from Midgard, inward toward the small, lies a living cellular world, and beyond that lies the quantum world. These microlands are the evolutionary and physical sources of everything we are. That may not make them gods, but compared to us they are more prolific, more ancient, universal, and omnipresent.

Midgard is not a place. It's a setting of the intellectual zoom lens. If you visited a planet in a galaxy a billion light-years away, your Midgard-scale intuition developed on Earth would be very useful, though fallible, on that planet. But if you sit comfortably in your chair at home and just change your mental focus to things far outside Midgard, in either direction, large or small, your intuition becomes completely useless. Without science, no one can—and no one ever did—imagine accurately how very small or very large things actually behave.

Only in Midgard do objects meet people's intuitive test of what's physical. "Physical" is actually an intuitive concept, not a clearly defined one. In common usage the word *physical* means solid, observable, and unquestionably here.

But outside Midgard, for example, the largest structures, superclusters of galaxies, are actually expanding apart faster and faster, and in billions of years they'll disperse. They're bound together not by gravity but by our dot-connecting minds, so in some sense they're not really physical objects. In the opposite direction from Midgard toward the very small, there are elementary particles that are not particles like marbles but rather quantum mechanical objects that are routinely in two or more places at once. What's "real" about them is the *probability* that something is happening. The strange truth is that what we usually think of as "physical" is a property of Midgard, perhaps the defining property, and thus Midgard is what people generally think of as the "physical" universe. Beyond Midgard, however, lies most of the Cosmic Uroboros. So most of the universe is not physical in the commonsense meaning of the word, but it is nevertheless real and governed by the laws of physics.

The ancient and medieval worlds believed that the ordinary world is somehow embedded in a spiritual reality; metaphorically this is an intriguing way to think about the location of Midgard on the Cosmic Uroboros. After all, those realms beyond Midgard are real, yet we experience them not directly but only through the intellect, the imagination, or perhaps an innate awareness that we are connected to the universe as a whole. This kind of spiritual realm is governed by the laws of physics—it's the kind of spiritual realm that could actually exist.

Until the twentieth century the universe was generally thought of as an endless extension of Midgard—for example, stars scattered outward forever, some with planets. But now we know that once you jump off the island of Midgard, nothing is the same.

Each size scale is in seamless continuity with the next, and yet every few powers of ten there's a *qualitative* change: a new phenomenon, like temperature or consciousness, emerges, or a hitherto irrelevant law of physics takes control. Choppiness in the way complexity grows is characteristic of the entire Cosmic Uroboros, and this can be stated as a law, which could be called the law of Uroboros thinking. The law states that *qualitative change every few powers of ten is a universal pattern.*

The law of Uroboros thinking applies to everything, including ourselves. After all, exponential growth happens not only in sizes of objects but also in the complexity of human interactions. This explains Friedrich Nietzsche's famous epigram, "Madness in individuals is something rare—but in groups, parties, nations, and epochs it is the rule." Everyone has probably noticed that although individuals can be kind, generous, and wise, a committee, corporation, country, or other type of group is almost never kind, generous, or wise even if most of the individuals who make them up are. An October 9, 2009, *New York Times* headline read, "Have Banks No Shame?" No, because banks can't feel shame. Shame is a human emotion, and a collective cannot *in principle* act or feel like a human being because it's not human, even though it's made of humans. This may be hard to believe at first, but think of it in reverse and it becomes clearer: you are entirely made of elementary particles, but you don't behave anything like elementary particles. You are many orders of magnitude more complex than elementary particles, and with your increasing complexity, new properties emerge and those of elementary particles disappear.

Every traditional ethical system teaches how to get along with other individuals in the same tribe, but now modern leaders, both political and corporate, control events on enormous time and size scales that no religious or political tradition has ever even contemplated. Big changes in thinking are required for decisions affecting such scales, but until people accept the law of Uroboros thinking they will not bother to figure out what those changes should be. Understanding that this is a universal law could help us accept it and begin to figure out how groups from the size scales of families to civilizations can collaborate more productively and respectfully with other groups. No religion teaches how to do this, at least not yet.

There are benefits available not only on the political and social scales but also individually to people who recognize their accurate and rightful place in the universe and adjust their thinking accordingly. To feel your roots stretching back through cosmic time is to *know who you are, and see from a cosmic perspective*. These are two abilities that our species needs to cultivate far more widely if we wish to protect the chances for our descendants over a cosmologically long period of time.

In later chapters we will come to the question of how to live in this new universe, after we have explained the new scientific understanding of time, how the universe began, what it's mostly made of, how we came to be here, and how we can create stories about these things that are powerful and bonding yet accurate. Science is the foundation of our reality, but discovering and expressing its human meaning is what will let the new cosmology have a positive impact on our lives. We create a cosmic society by expanding our thinking to encompass what we now know exists, by expanding our sense of identity to take our true place in it, and by looking from this new perspective at our behavior on all size scales.

Chapter 3
We Are Stardust

What are we human beings made of—literally? Flesh and blood? Mind and body? To become a cosmic society, we need to appreciate at a much deeper level what "we" are and how this larger we fits into the universe. Our bodies are made of many kinds of complex atoms, most of which were created inside ancient stars or during supernovas and then flung out during the violent deaths of those stars to travel for eons through space. We are 90 percent stardust by weight and 10 percent hydrogen (mostly in our H_2O). We and the ground we walk on are literally made of stardust.[1]

But isn't everything? Well into the twentieth century, scientists believed that everything is made of atoms—and everything on Earth actually is. But now we know that Earth is extremely atypical of the universe as a whole, and the way things operate here is not a good basis for extrapolation. Most of the universe is not made of atoms. Until astronomers discovered this, we couldn't begin to comprehend how we humans fit into the whole, because despite having a word for the whole—*universe*—we had no idea what that really was. But now we know that stardust is the rarest material in the universe and exists only because the conditions for its creation were set up by the universe's two overwhelmingly dominant ingredients: invisible dark matter and invisible dark energy.

The Pyramid of All Visible Matter represents everything visible in the universe that can be detected with any scientific instrument (fig. 23). We borrowed the symbol from the back of the dollar bill and the Great Seal of the United States. The thirteen brick steps originally represented the thirteen colonies, and the Eye represented the hope that Providence would look kindly on this historic undertaking. The ribbon at the bottom says in Latin, "The New Order of the Ages." This motto originally referred to the founding of the new country with its new kind of government. We're radically reinterpreting this entire symbol, but the motto is even truer of the new universe. In the Pyramid of All Visible Matter, the volume of each section of the pyramid is proportional to the amount of the corresponding ingredient in the visible universe. Hydrogen and helium came straight out of the Big Bang and fill the entire bottom section of the pyramid. They're the lightest atoms that exist, but there are so many of them in space that collectively they vastly outweigh all the other atoms. Hydrogen and helium eventually condensed and ignited, creating the first stars. Inside the stars, nuclear processes created new kinds of atoms that the young universe had never before seen, such as carbon, oxygen, silicon, and iron. These heavy atoms (heavier than hydrogen and helium) are ejected from stars at the ends of their lives and become stardust, floating through interstellar space, perhaps to be pulled into the gravitational field of some newly forming

Fig. 23. The Pyramid of All Visible Matter

star system, making it possible for such later-generation stars to have rocky planets like Earth.

The periodic table of the elements is familiar from any high school chemistry classroom, but here it is color-coded to show the origin of each element (fig. 24). Hydrogen and helium, alone at the top, are in some sense the parents of all the others, and thus the great-great-

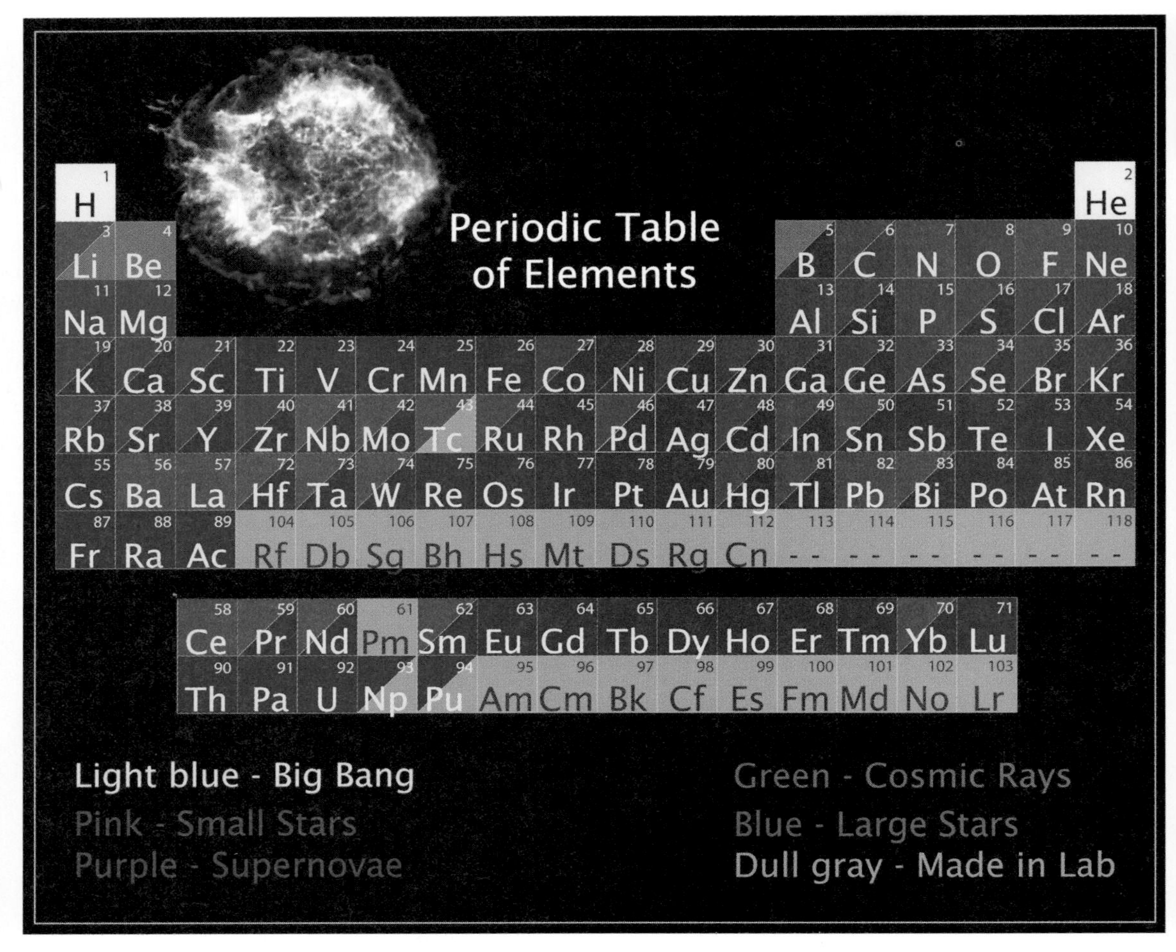

Fig. 24. The periodic table of the elements, with the origins of each element

grandparents of us all. The iron atoms in our blood carrying oxygen at this very moment to our cells came largely from exploding white dwarf stars (fig. 25), while the oxygen itself came mainly from massive stars exploding as supernovas (fig. 26).

Most of the carbon in the carbon dioxide we exhale on every breath came from planetary nebulas, which are the death clouds of middle-size stars like the sun. The Cat's Eye Nebula is an extraordinarily beautiful planetary nebula. Here it is shown among its neighboring stars (fig. 27).

▣ When we zoom in, we see what looks like an eye at the center (fig. 28). All that's left of the original star is the tiny white dot at the center of this close-up. The rest of the star has blown off in those colorful clouds of stardust that surround the star in multiple layers and may someday become part of a new planetary system.

Distant worlds may be wildly different from Earth, but there are things that must be true of them all, simply because of the nature of stardust. For example, on any planet in the Galaxy, wherever you find watery seas and land, there will be sandy beaches (fig. 29). This is because oxygen and silicon are two of the most abundant heavy atoms produced before a star explodes in a supernova. Free-floating in space, they combine with each other and the hydrogen that is everywhere, making H_2O and SiO_2—water and sand—which travel together and become incorporated into new worlds.

It takes a star millions or billions of years to produce a comparatively tiny number of heavy atoms, yet heavy atoms incarnate our world. The capstone of the Pyramid of All Visible Matter—the floating part at the top—represents the total mass of stardust compared to that of hydrogen and helium. The Eye in the capstone is the reason we chose

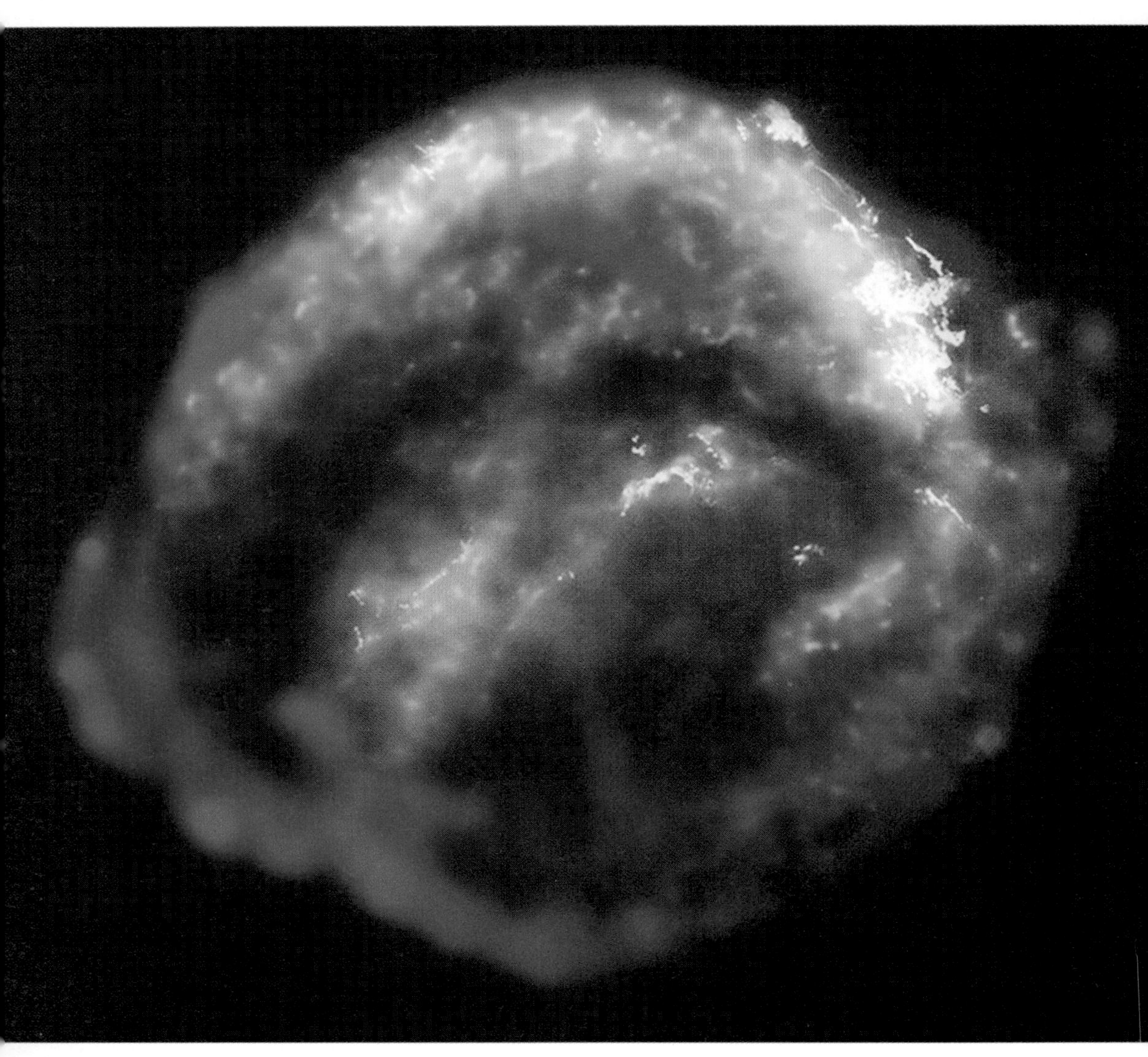

Fig. 25. Kepler's supernova remnant, from the explosion of a white dwarf

Fig. 26. The Crab Nebula, remnant of the explosion of a massive star

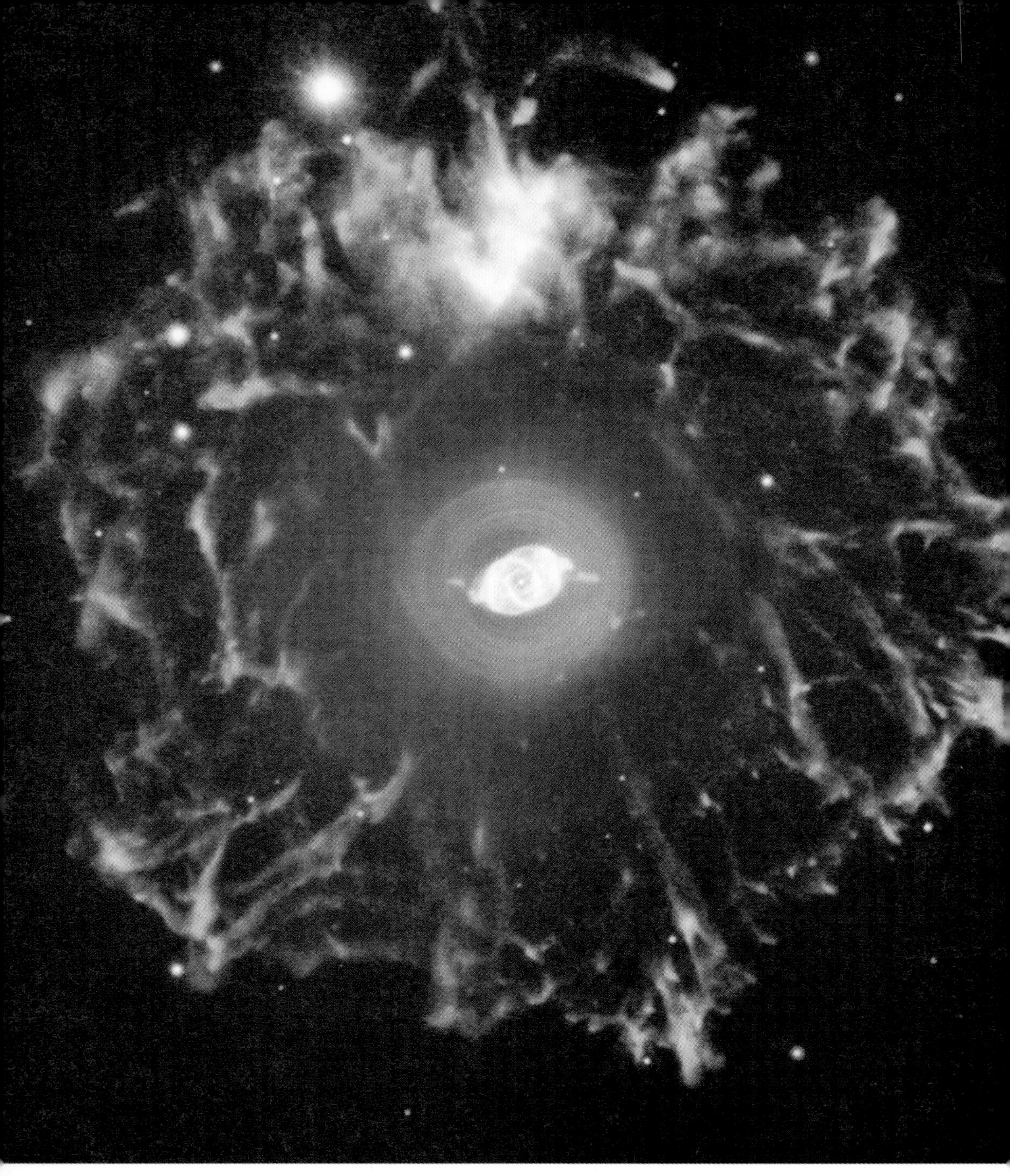

Fig. 27. A large-scale view of the Cat's Eye Nebula

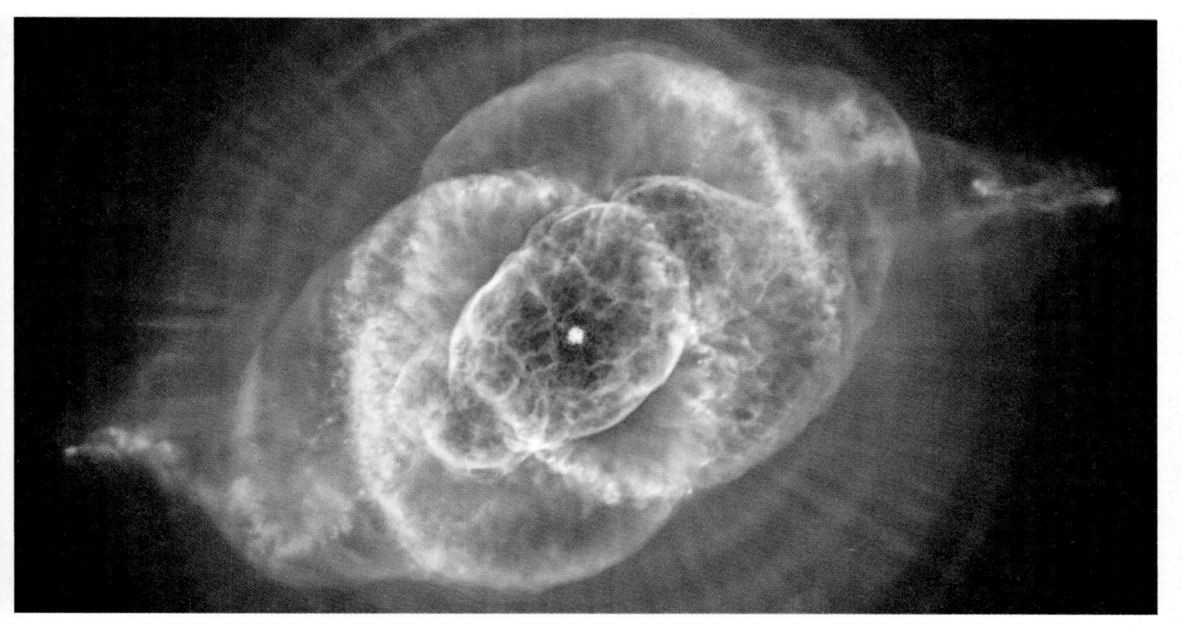

Fig. 28. The Cat's Eye Nebula

Fig. 29. Manorbier Beach, Pembrokeshire, Wales

this symbol (rather than, say, a food pyramid or a pie chart). The Eye represents the minuscule amount of stardust in intelligent creatures—on all worlds anywhere. This is the only cosmic ingredient not drawn to scale, since if the Eye were to scale, it would be a microscopic point.

The Pyramid of All Visible Matter represents what was once thought of as the whole universe. Midgard on the Cosmic Uroboros was also once thought of as the whole universe, but in both cases our perspective has since expanded. We now know scientifically that visible matter is only a small part of the new universe.

The Cosmic Density Pyramid represents everything—visible and invisible—that gives the universe density (fig. 30).[2] This includes not only matter but energy, which, as Albert Einstein showed with his iconic equation, $E = mc^2$, is convertible to mass. The Pyramid of All Visible Matter is just the aboveground tip of this huge underground pyramid.

In addition to the half a percent of atoms that are visible, the universe includes an extra 4 percent of atoms that are invisible simply because they are floating through space between the galaxies, far from any stars and thus unlit. The two dominant ingredients in the cosmic recipe—the invisible 95 percent of the density of the cosmos—are dark matter and dark energy. Both are unearthly substances whose natures are still not well understood, although thousands of scientists are working on it.

Dark Matter and Dark Energy

Most of the matter holding the Milky Way and all other galaxies together is cold dark matter. Dark matter is invisible not because it is unlit but because it does not interact with light at all. Dark matter does not emit light as stars do, reflect light as planets, moons, and gas clouds do, or

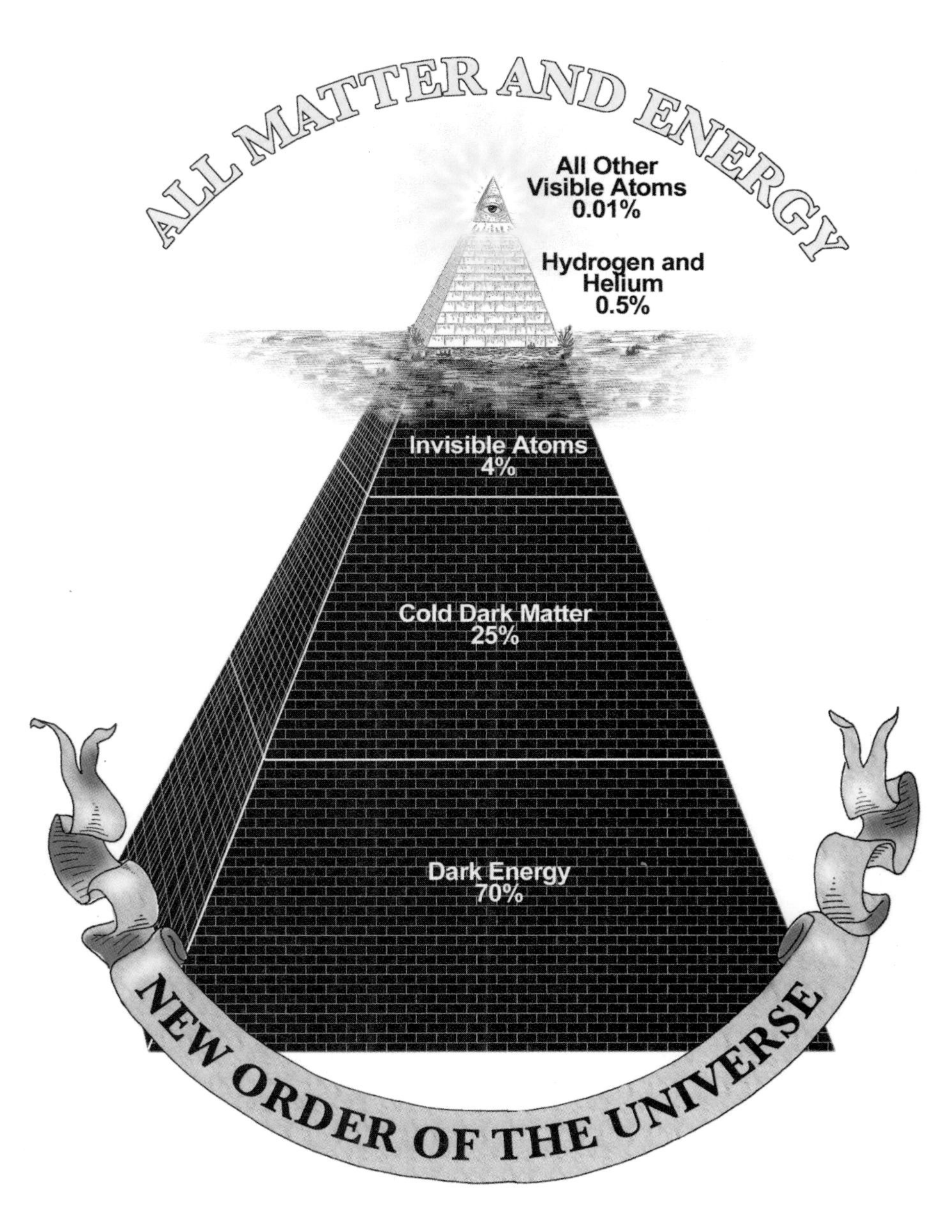

Fig. 30. The Cosmic Density Pyramid

absorb light as dust does. It also doesn't emit or absorb X-rays, radio waves, or any other form of radiation that astronomers have detected. We know dark matter is there only because of its immense gravitation, which affects the objects around it in ways we can measure. Because dark matter hardly interacts with anything, including itself, it can never evolve into anything complex. It just forms big "halos"—blobs of eternally whizzing dark matter particles that surround and permeate each and every galaxy. Dark matter is not a good name, in fact, since it's not dark: it's transparent. But whatever it's called, it controls the origin and evolution of galaxies, galaxy clusters, and superclusters of galaxies because it concentrates the matter and helps determine how everything else in the galaxy moves.

The largest portion of the Cosmic Density Pyramid, about 70 percent of the density of the universe, is dark energy. This is the most powerful entity in the universe, and yet until 1998 no one knew it existed. Scientists thought of it only as a hypothetical possibility. Dark energy powers the expansion of the universe, and that expansion is a key part of understanding the picture of our universe. Here's how it works.

All distant galaxies are being carried away from our Milky Way by expanding space itself. The galaxies are not flying away from one another across space; the space between them is expanding, and the farther away from us, the faster it's expanding. If we look at two galaxies, and one is twice as far away from us as the other, then from our perspective the farther one will be receding twice as fast. Every observer in the universe sees exactly the same pattern in the motion of distant galaxies. That's how a uniformly expanding universe works.

For much of the twentieth century astronomers assumed that

the expansion of the universe must be gradually slowing because of the mutual gravitational attraction of everything in it. But in 1998 the amazing discovery was made that the expansion of the universe is not slowing down at all but is instead accelerating. Dark energy makes space repel itself. The more space there is (and increasing amounts of space are inevitable in an expanding universe), the more repulsion there is. The more repulsion, the faster space expands. The faster it expands, the more space, the more repulsion, and this leads to an exponentially increasing expansion, which may possibly go on forever. Dark energy seems to be a property of space itself.

Imagine that the entire universe is an ocean of dark energy. On that ocean there sail billions of ghostly ships, made of dark matter. At the tips of the tallest masts of the largest ships there are tiny beacons of light, which we call galaxies (fig. 31). With the Hubble Space Telescope, the beacons are all we see. We don't see the ships, we don't see the ocean—but we know they're there through theory, the Double Dark theory. This oceanic imagery has a certain resonance with the ancient Egyptian and biblical cosmologies. Both of them envisioned that the primeval water, an unearthly substance, not normal water, predated the creation and continued to surround it. The world was created in the midst of it. But here's the twist in our modern version: dark *matter* is indeed primeval and came out of the Big Bang, and it surrounds and permeates our Local Group of galaxies, but dark *energy,* which dominates outside our halo of dark matter, only became important later, because it has been creating more of itself all the time. The larger the universe expands, the faster more dark energy gets created.

Until the late 1990s astronomy students were usually taught that there are exactly three possible futures for the universe:

Fig. 31. Dark matter ships on an ocean of dark energy

1. It can expand forever at a constant rate;
2. It can slow down forever asymptotically (that is, the rate of expansion would get closer and closer to zero without ever reaching it); or
3. With enough matter in the universe, the gravitation of matter could eventually slow the expansion to a halt and the universe would begin to contract. This last possibility led people to speculate that everything might come back together again in a Big Crunch, ending the universe symmetrically, and then maybe there would be another Big Bang, ad infinitum.

We bring all this up because these alternatives have been discussed so widely. But all three are now known to be wrong: the expansion of the universe is actually accelerating. How do we "know" this? What's the evidence?

The real test of a theory is not how logical, beautiful, and satisfying or, on the other hand, how weird, ugly, or unlikely it sounds. The test is, how accurate are its predictions?

The Double Dark theory has made many precise predictions about phenomena that had never been observed or in some cases even looked for. Over the past twenty years huge amounts of data have become available from the Hubble Space Telescope, from powerful new telescopes on the ground, like the Keck Observatory in Hawaii, and from satellites observing not visible light but the heat radiation of the Big Bang. The predictions of the theory have turned out to be right without exception.[3]

In figure 32, the wavy blue line that looks like a mountain range is the Double Dark theory's prediction that a specific pattern in the heat radiation from the Big Bang would be discovered.

It is not necessary to master most of the graph in figure 32; its sole

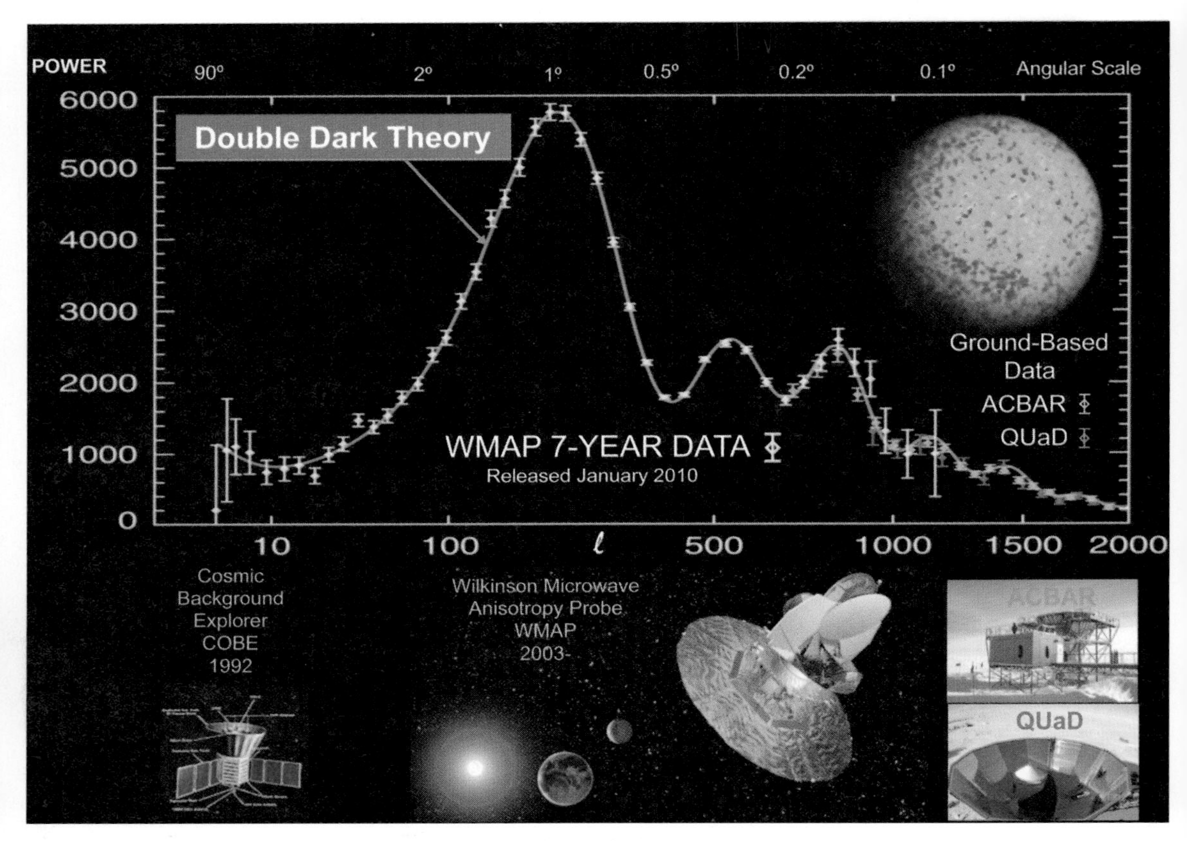

Fig. 32. Big Bang data agree with the Double Dark theory

purpose here is to illustrate the extraordinarily close match between the predictions of the theory and the observational data. This complex prediction was made before there were any data and before there was even an instrument capable of making the observations. Every white point is a later separate observation. Over the years, as better measurements have been made, every single observation has fallen along the predicted curve. This can't be a coincidence.

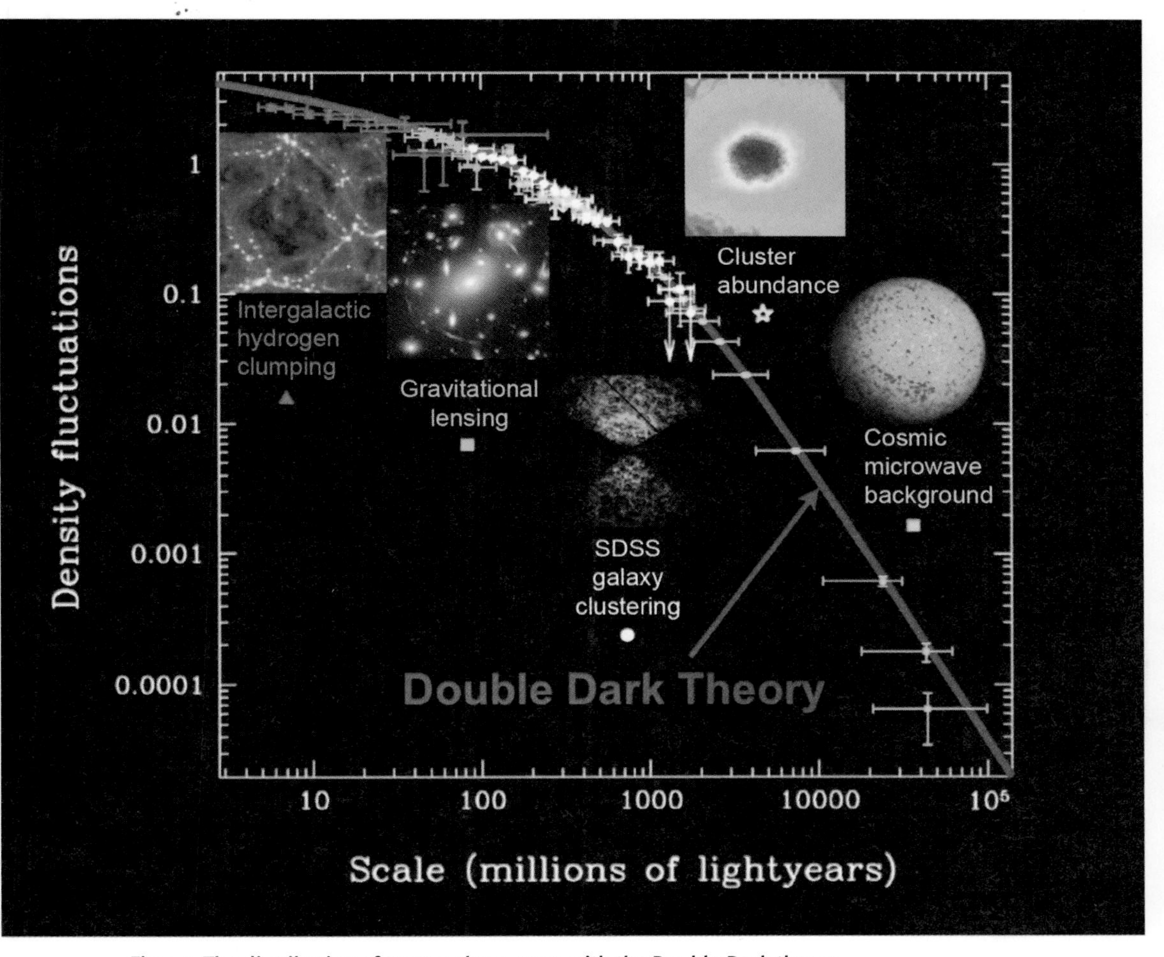

Fig. 33. The distribution of matter also agrees with the Double Dark theory

In figure 33 the red curve was the Double Dark theory's prediction of how matter should be distributed in the universe today on all size scales from our local galactic neighborhood out to the cosmic horizon. Once again as the data have come in, every point has fallen on the line.

Hidden in these graphs is the real story of the expansion of our universe. In the early stages of the universe there was the same amount of dark matter as there is now, but there was relatively little dark energy because there was relatively little space—the universe hadn't had time to expand very much. And so, for the first nine billion years the gravitational attraction of the dark matter did in fact slow down the expansion. But the expansion kept going, producing always more space, so eventually the dark matter thinned out. Dark energy, however, doesn't thin out, perhaps because it's a property of space. Its relative importance only increases as the universe expands. Now the repulsive effect of the dark energy has surpassed the gravitational attraction of dark matter as the dominant effect on large scales in the universe. The turning point was about five billion years ago—which was, coincidentally, about when our solar system was forming.

Wild Space versus Tame Space

Taking the existence of dark energy seriously means we have to begin thinking in a new way about what "outer space" means. Most people use the phrase to refer to any place outside Earth's atmosphere, but this communicates a static picture of the universe—it's missing the whole idea of expansion.

If we think of outer space as starting outside Earth's atmosphere, then we need to think of it as *stopping* at the edge of our "Local Group" of galaxies (the Milky Way, Andromeda, and their fifty or so satellite galaxies). Because our Local Group is bound together by gravity, it travels as a unit in the great expansion. It doesn't expand apart. There's no expansion happening to people, planets, or even the entire Galaxy, nor between our Galaxy and the rest of the Local Group (fig. 34).

Fig. 34. "Don't Feel Bad Loretta . . . The Entire Universe Is Expanding"

Nice excuse, Lockhorns, but sorry. Gravitationally bound clumps like our Local Group are actually contracting and falling together and will merge in a few billion years. But *outside* the Local Group, space is expanding faster and faster.

So all the space inside the Local Group is a special kind of space, because it's been tamed by gravity. Outside our Local Group lies the real outer space: wild space. On this huge scale, dark energy is tearing apart all large structures and accelerating the rate of expansion. Wild space is carrying hundreds of billions of galaxies away from our Local Group in all directions.

To show how wild and tame space work, we can visualize the dark matter expanding, and we do this below with a sequence from a computer simulation. Whenever we simulate dark matter, brightness is used to represent density: the brighter a region appears, the more dark matter it contains, although in reality dark matter is completely invisible.

▣ In figures 35–37 we take a small portion of the universe and show it expanding.

Quickly the screen becomes filled with a region that has become bound together by gravity, and that region stops expanding and instead grows by attracting other blobs of dark matter (fig. 35). After about seven billion years (fig. 36), the central region of this dark matter halo has stopped expanding, and some of the dark matter halos around it are starting to fall toward it. But some don't. In figure 37 the small dark matter halo in the upper left corner of figure 36 is continuing to expand away from the central halo—so the space between that halo and the central halo of figure 36 is wild space.

Yet expansion is not all that's going on. Regions with slightly more dark matter than average expand slightly more slowly than regions with less dark matter. The "rich" regions (rich in dark matter) get richer and the poor regions get poorer until the differences become substantial. Inequalities that were there in the beginning get magnified. When a region has twice as much dark matter as average, it stops expanding, while the lower density regions around it keep expanding. Once a galaxy-sized region stops expanding, the ordinary matter in it can fall to the center and begin to form stars. The dark matter is forming structures on every large size scale, and the sum of them all is called the *cosmic web.*[4]

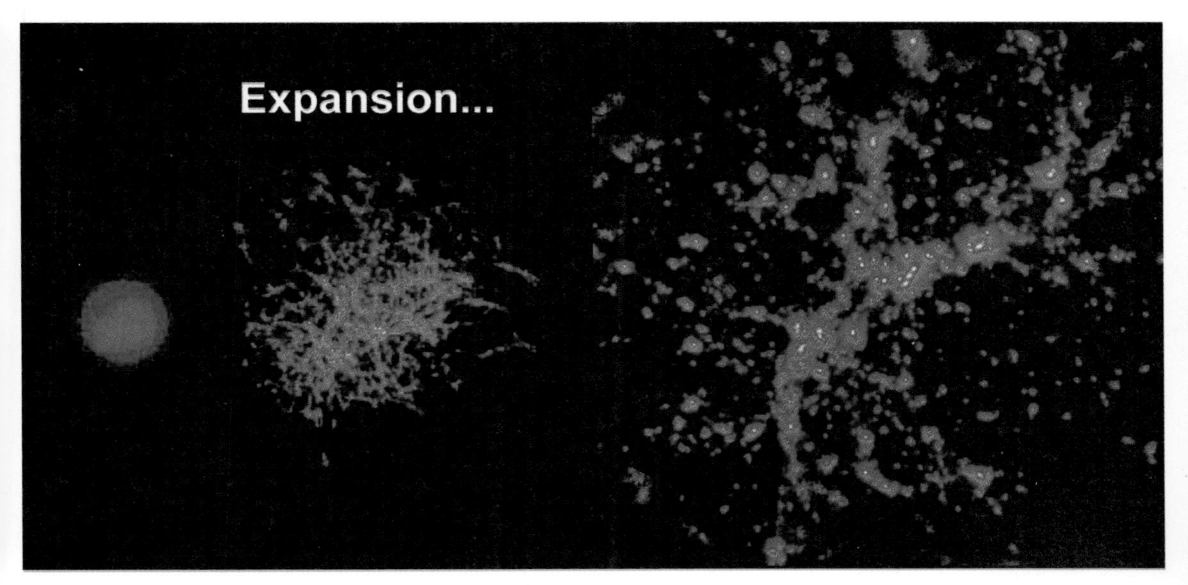

Fig. 35. A simulation of the expansion of the universe

Suppose that a filmmaker wants to show a scene between two characters inside a dining car on a moving train. If she or he set up the camera outside on the ground and tried to film the scene through a window as the train went by, the camera would catch only a second of the action. Instead the filmmaker films from inside the dining car, thus subtracting out the movement of the train. We're essentially doing the same in the sequence in figure 38. We want to show how the dark matter is forming structures, not how those structures are simultaneously getting stretched by expanding space, so here the expansion of the universe is subtracted out. In the sequence we have also blown up all the earlier time steps to the final size so that we can focus on what is happening inside. Astronomers call this "working in co-moving coordinates."

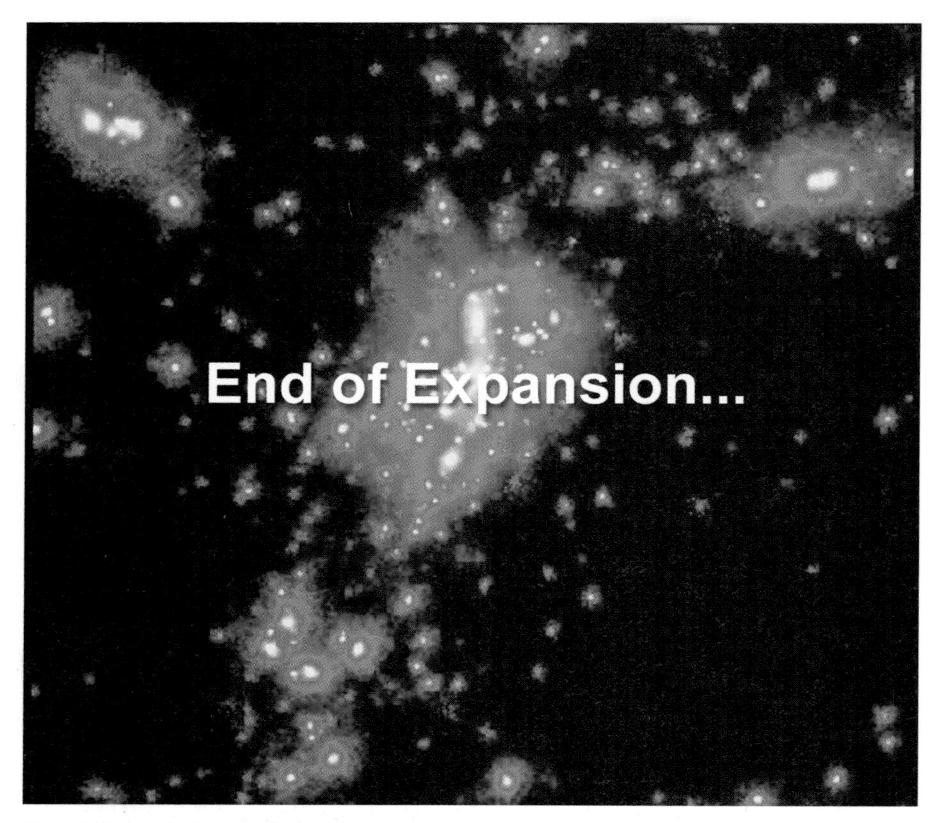

Fig. 36. The end of expansion

In the first box, which represents a cube of the universe shortly after the Big Bang, the dark matter starts out very smooth. But gradually the dark matter is being attracted to wrinkles in space-time (this process is explained in chapter 5) and begins to show some structure. As time passes, the structure becomes sharper: the filaments and intersections of filaments become denser and the voids in between become emptier. Galaxies form inside the filaments; inside the intersections of

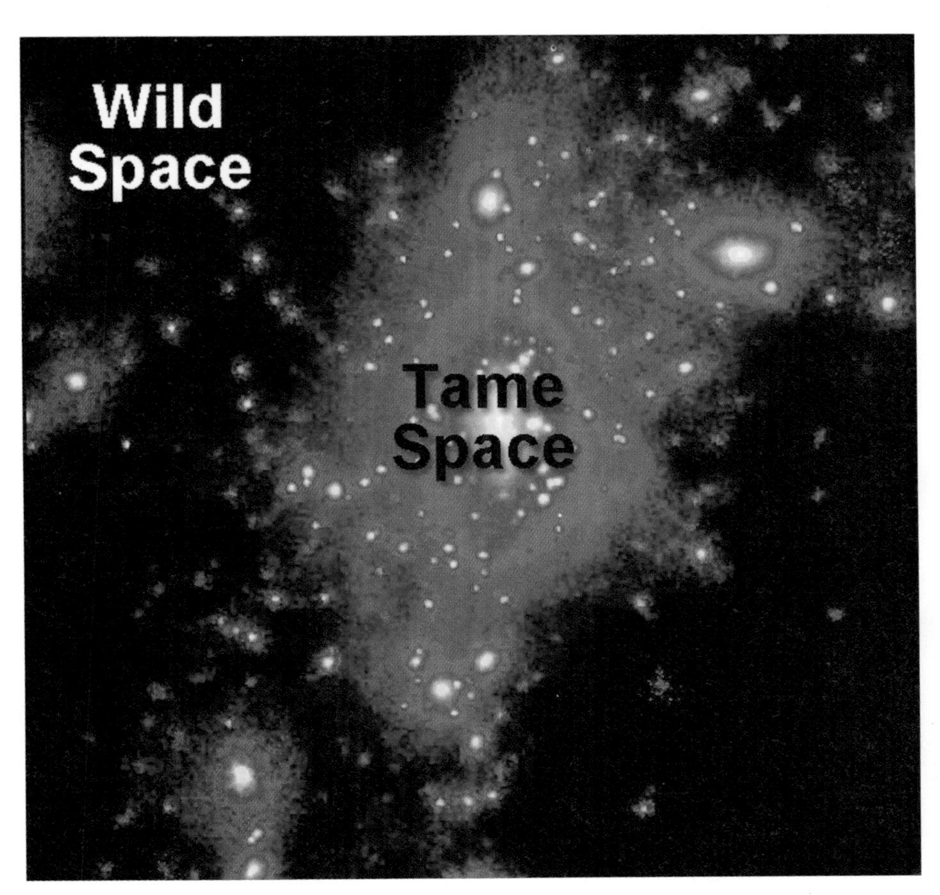

Fig. 37. Wild space, tame space

the filaments, where the dark matter is densest, clusters of galaxies form ▣.

The highest-resolution large-scale simulation of the evolution of the Double Dark universe yet run is called Bolshoi and is a pathbreaking new supercomputer simulation that lets us see the invisible dark matter in unprecedented detail (fig. 39). In it every single dark matter

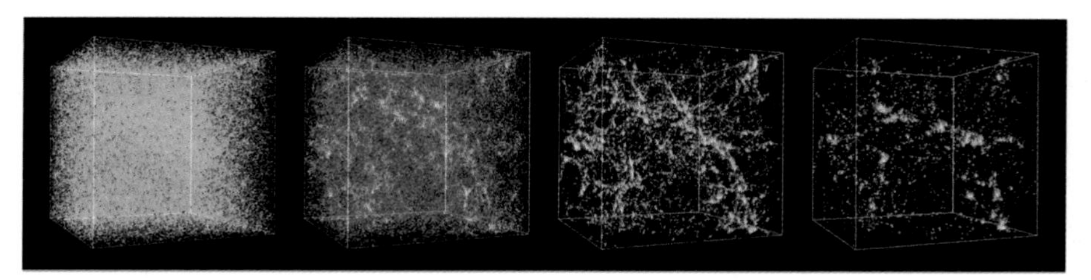

Fig. 38. The evolution of the cosmic web

halo that could host a visible galaxy is shown. The region simulated is about a billion light-years across. The distribution of dark matter halos in these simulations looks statistically just like the actual distribution of galaxies in the universe ▣ ▣.[5]

In figure 40 the magnificent Aquarius simulation ▣ shows the formation of a single Milky Way–sized dark matter halo. How big would the dark matter halo be, compared to the final visible Milky Way? The tiny photo in the middle would be the visible part of the Milky Way. The halo is so huge that it engulfs several smaller galaxies.

There is a connection between events on this immense size scale and our own lives. Dark matter doesn't cradle the entire Milky Way—and all galaxies—in delicate, invisible hands, protecting it from the cosmic hurricane of dark energy tearing space apart outside because it cares about us. Dark matter didn't herd a dispersed, fertile mix of hydrogen and helium into a compact region at the center of our Galaxy so that those primal atoms could easily interact and evolve into stars and worlds in preparation for us. Dark matter didn't commit itself unconditionally for billions of years, with never an instant off, to hold the atoms in its charge safely in a stable home galaxy for us. It does all these things because it has no choice. Its behavior is built into the order

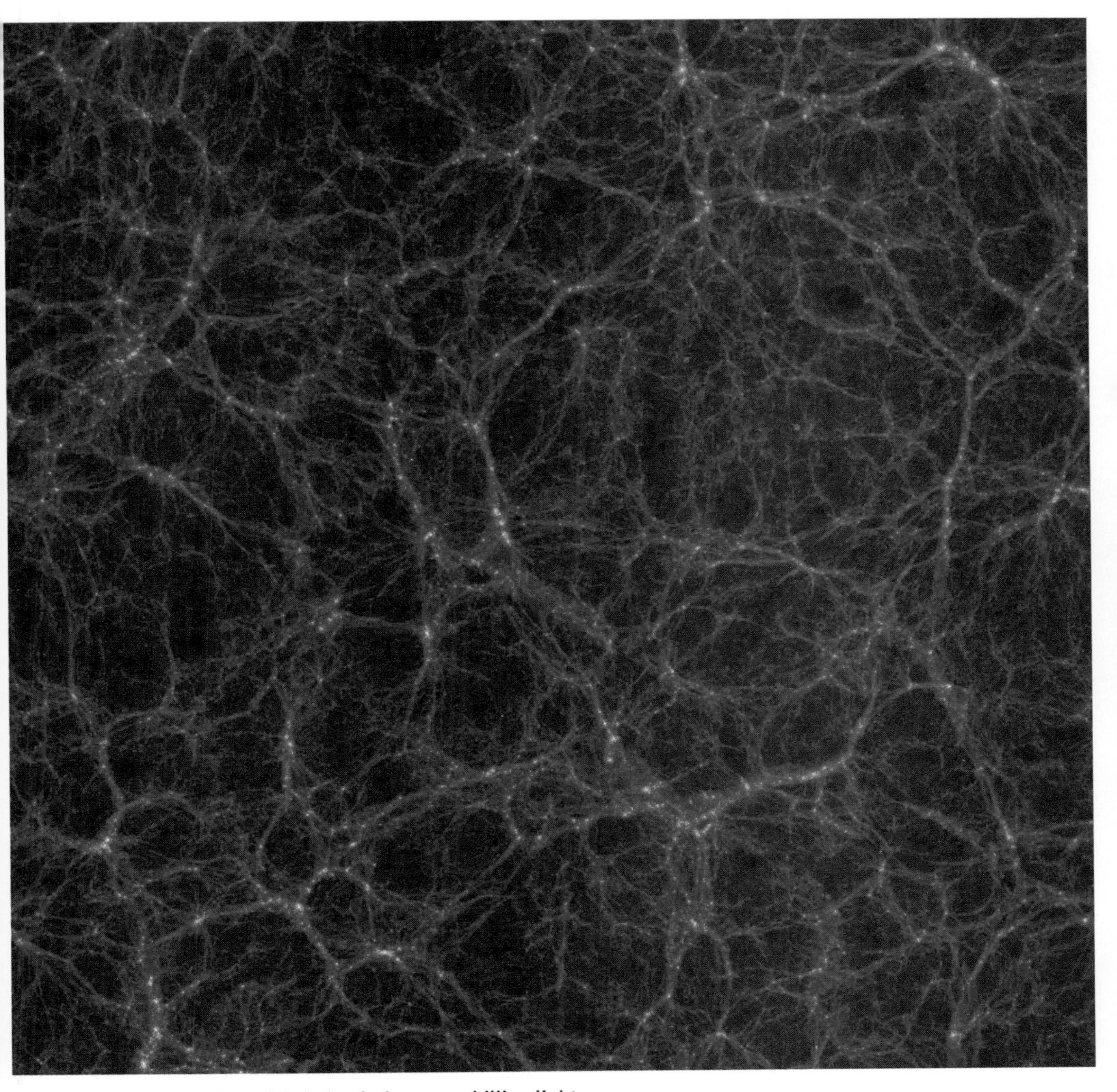

Fig. 39. The Bolshoi simulation—one billion light-years across

Fig. 40. The Aquarius simulation of a Milky Way–size dark matter halo

of the universe. But we benefit. We minuscule bits of stardust in that tiny Eye at the peak of the Cosmic Density Pyramid are the only ones in this universe who realize and can appreciate the immensity of what dark matter does and has done.

And yet, beyond shaping and protecting the galaxies, dark matter has evolved very little; it has not become more complex, and the reason is that it doesn't participate in chemistry. Its particles scarcely interact with one another or anything else, except gravitationally. Chemistry is created by myriad electromagnetic interactions among atoms and is incomparably more complex than physics. Biology is more complex than

Fig. 41. The Eye of the Pyramid of All Visible Matter

chemistry, and we civilized, intelligent beings are the most complex things we know of in the entire universe.

We, with the other potential members of the cosmic club of intelligent life, are at the peak of the Cosmic Density Pyramid. An enormous base of material and phenomena in the universe has made and will continue to make our existence possible. Within the floating capstone, the fraction of stardust associated just with living things or the remains of living things is extremely tiny. Within that extremely tiny fraction, the fraction associated specifically with intelligent life anywhere in the universe is vanishingly small—yet it is only *that* which looks at and grasps this pyramid and the way that time has constructed it. As much as people around the world hope to find alien beings on other planets, the possibility exists that only our eyes see this universe (fig. 41).

Intelligence can burst out only from bits of stardust. Everything we learn about ourselves in the context of the universe as a whole reinforces a fundamental fact: that from a cosmic point of view we intelligent, self-reflective beings are rare and precious beyond calculation—but we are only possible because of the composition of the rest of the universe.

Chapter 4
Our Place in Time

Time is as much a part of us as the matter we're made of. People tend to think of physical objects, including our bodies, as things that fully exist here and now, although they have a history. But this is an old-fashioned view of time. Time exists on many different scales, and by understanding this we begin to see that we self-reflective beings also exist in different ways on different timescales.

To create the famous photograph called the "Hubble Ultra Deep Field," Hubble Space Telescope stared for more than two weeks in one direction into what looked from the ground like empty, black sky (fig. 42). If you held two sewing needles at arm's length and crossed them, their intersection would be the size of the patch of sky visible in this photograph. After collecting two weeks of light, Hubble saw hidden in the darkness all these faint and distant galaxies. This was the first time people had seen them in great detail. The colors here are as the human eye would see them, if we could see such faint light. (The Hubble Ultra

Fig. 42. The Hubble Ultra Deep Field in optical light

Deep Field image in chapter 1 shows the same region of sky but in infrared light, which is invisible to the human eye.) This seems to be a still shot, not showing time, but time is in fact clearly in the photograph. The light coming from each of those galaxies has been changed by time—it's been stretched (redshifted) by the expanding space it has crossed, and the more space, the more stretched it becomes. A stretched wavelength shifts the color of the light toward the red end of the spectrum, and the amount of redshift tells us how much the light has expanded with the universe since it was emitted. And this in turn tells us in what era of the universe's evolution we are seeing the galaxy it came from. We're seeing that galaxy exactly as it was when the light left it, possibly billions of years ago—not as it would look today to an observer nearby it. We will never see what it looks like today, except by waiting billions of years for the light that's leaving it now to arrive here, but we can calculate *where* it is now and how far back in time we are seeing it.

▣ Using this information, we can spread the galaxies out in three dimensions and zoom into the photograph. *As we zoom in, we travel back in time.* In figure 43 the relatively nearby galaxies disappear, and in figure 44 more distant ones have disappeared. Nearby galaxies are often big disk galaxies like the Milky Way or elliptical galaxies (big round balls of stars with no disk). But as we travel back in time, the nature of the galaxies changes. When we get back ten billion years, we see no big galaxies like those near us today. In the first two billion years after the Big Bang, galaxies are small and irregularly shaped but very bright because they are making stars very rapidly (fig. 45). These galaxies are ragged looking, probably because at these very early times the universe was a lot smaller than it is now but had just as much dark mat-

Fig. 43. Zooming in to the Hubble Ultra Deep Field

Fig. 44. Looking back to the first three billion years in the Hubble Ultra Deep Field

Fig. 45. Looking back to the first billion years in the Hubble Ultra Deep Field

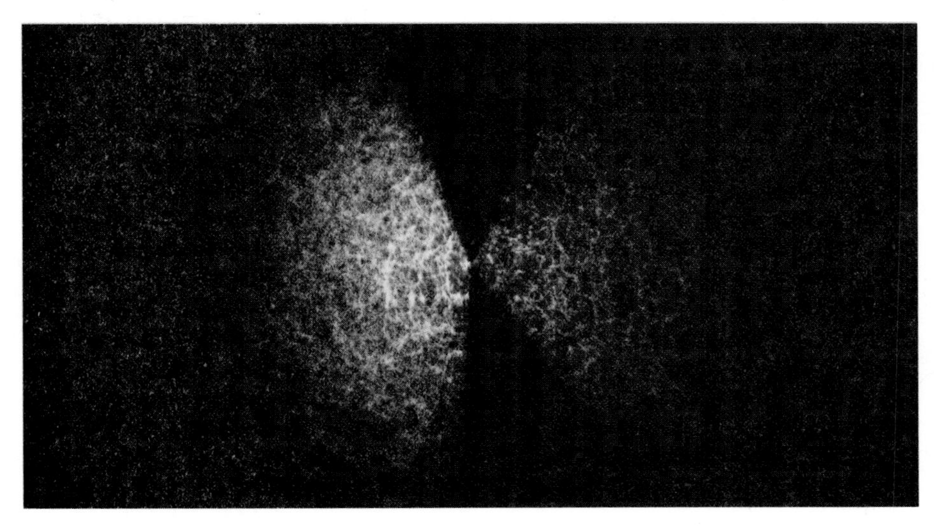

Fig. 46. The Sloan Digital Sky Survey

ter, so the dark matter halo of a forming galaxy was frequently colliding with the dark matter halos of other galaxies. If we go back far enough, there are almost no bright galaxies, because they haven't formed yet.

The Hubble Ultra Deep Field captured the depth of just one point on the sky. We now know the location of far more galaxies than these. About a million galaxies have been mapped by a huge ongoing project called the Sloan Digital Sky Survey, whose goal is to discover the large scale distribution of galaxies.

▣ Whole slices of the universe have been mapped all the way out to the Big Bang in directions visible from the mapping telescope (fig. 46).

The sphere of cosmic background radiation, the heat radiation from the Big Bang that fills the universe, is shown in figure 47. Deep inside the universe the Sloan Digital Sky Survey map of the galaxies is visible in white. The colors in figure 47 represent slightly different tem-

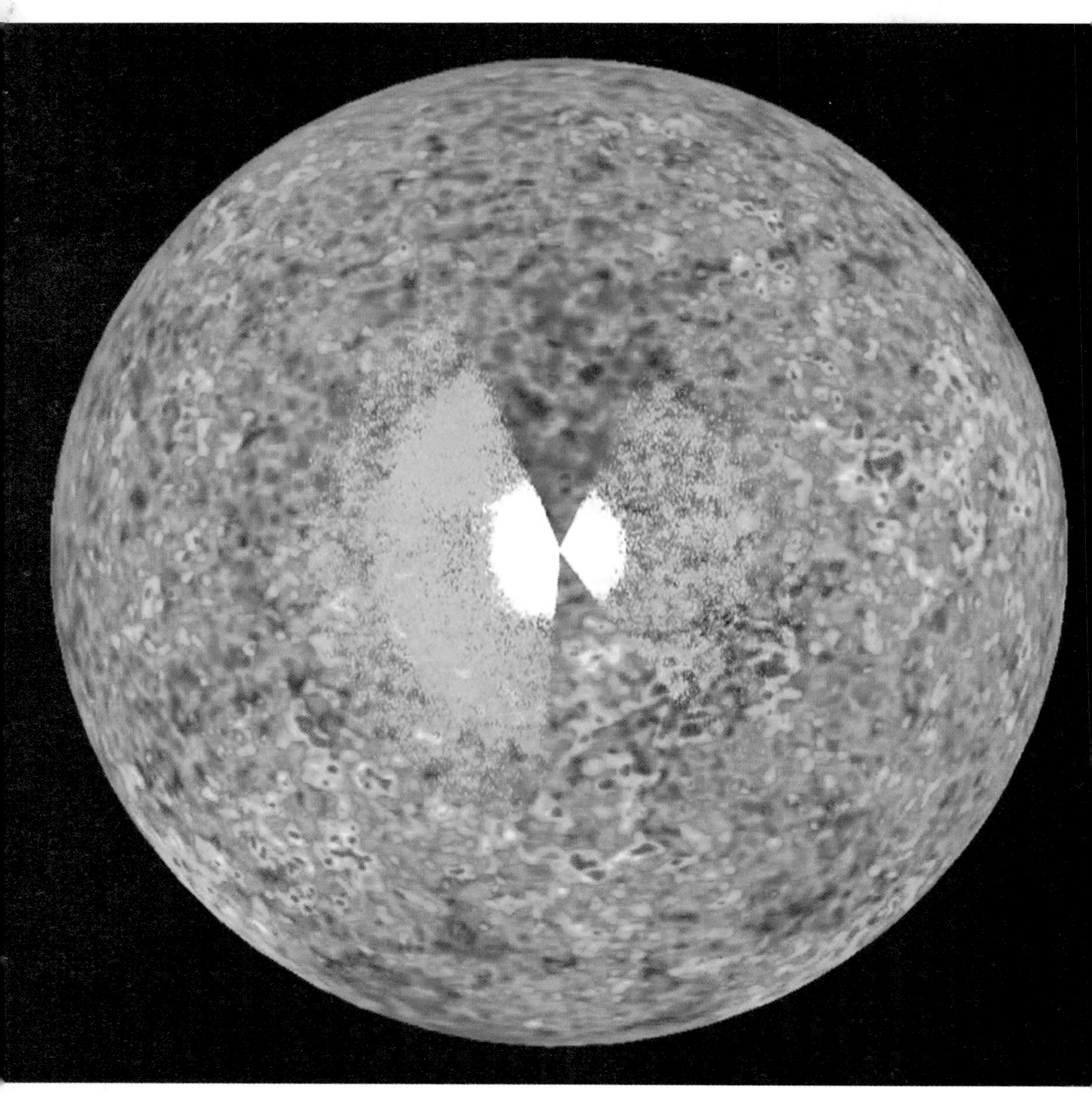

Fig. 47. The cosmic microwave background radiation sphere

peratures in different directions, and those variations in temperature today reflect the slight differences in density when the universe was only four hundred thousand years old.

But there's something weird about figure 47: the sphere is shown from the outside, as though we're outside the universe, looking at it, and not a part of it. To the extent that this is a helpful stance for scientists who are gathering or analyzing data, this imagined standpoint serves a purpose. But it's extremely misleading as a mental picture of reality. We modern people—both scientists and nonscientists—who unlike the ancients lack the intuitive sense of truly belonging to our universe, need to start visualizing our universe from the *inside,* where we actually are. Otherwise, we will always misinterpret ourselves, feeling as though we are outside, insignificant, sensing a familiar existential isolation, and looking at a universe in which we seem to play no part. But this is everyone's universe. No scientific explanation that portrays us as objective observers of a universe we're not part of can ever be satisfying. And more than that, no such explanation can ever be accurate, since the universe exists on all size scales, and on our scale we are indisputably part of the universe and must be understood in context.

Where in the evolution of the universe are we? One way to answer this question is with a picture. We call figure 48 the Cosmic Spheres of Time, and it represents our visible universe from the point of view of time. Our Galaxy is at the center, which represents Today.

When we look out into space, we are looking back in time. In the Middle Ages, Europeans believed that the earth was the center of the universe and was surrounded by concentric heavenly spheres, as we discussed in chapter 1. We have borrowed that medieval image to cre-

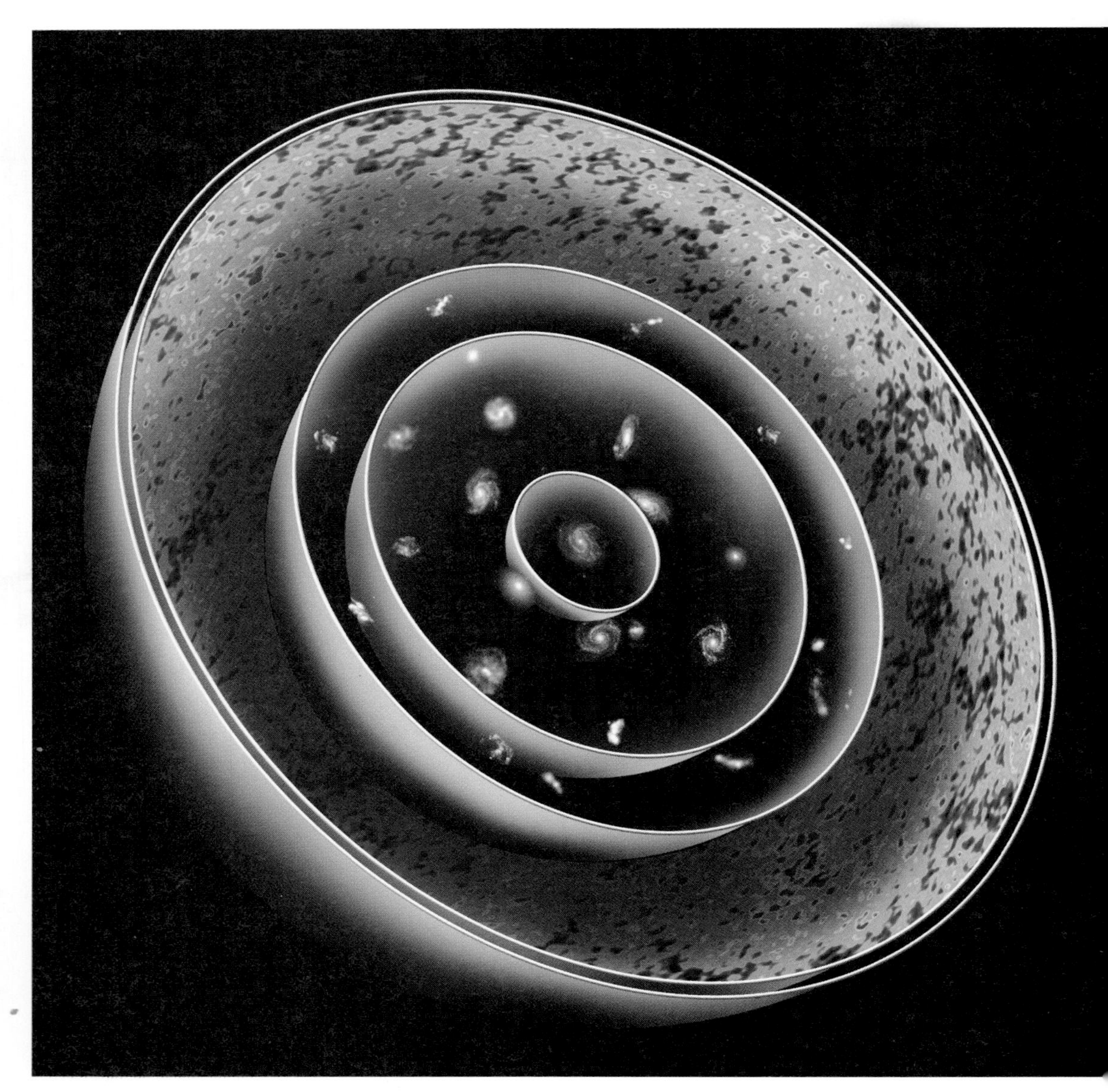

Fig. 48. The Cosmic Spheres of Time

ate this symbol, because humans do in fact always see outward from the center. But it's not the earth that is the center of the universe: our Galaxy *here and now* is the center of our *visible* universe. Every galaxy is the center of its own, unique visible universe. In this symbol, the distances between the concentric spheres don't represent just spatial distance—each sphere moving outward represents an earlier and earlier *time* in the history of the universe. The earliest time, the Big Bang, is represented by the outermost sphere.

The *era* when our sun and earth were forming, four and a half billion years ago, is still out there, spherically enfolding our solar system, our Galaxy, and all the nearby superclusters of galaxies. That era is represented by the innermost sphere. Far beyond that is the sphere representing the era when the first big galaxies formed. Far, far beyond, the era of the earliest bright galaxies engulfs us. Beyond that is a deep sphere of utter darkness that theory tells us is the real Dark Ages of our universe before the first galaxy ever formed. Earlier still lies the colorful sphere of the cosmic background radiation, which we are receiving from all directions in the form of microwaves. Beyond that lies the cosmic horizon, or Big Bang.

The Cosmic Spheres of Time place our Galaxy at the center, but it's a limitation of this two-dimensional picture that the spheres are drawn on the page while we are out here, looking at them. Jump in! In your imagination take your place in the center of the symbol, at Today, and then close the spheres around yourself. You are physically immersed in the past of the universe. Take a moment to absorb that. We are at the center of our past. The past is not "over": it's racing away from us at the speed of light like ripples from a pebble thrown into a pond, but not in circles—in spheres. Spheres of time.

Our human place in the cosmos is not a geographical center but a symbolic, meaningful place, created by the interaction of space, time, light, and consciousness. Without consciousness, after all, there would be no visible universe. Something would exist, but nothing can be said about it. The Cosmic Spheres of Time are real because the past is real. Light from far away and long ago carries the evidence of the past to us.

It may help to think of it this way. In the ancient world, when people needed to get a message to someone, they had no instant media of course; they had to send a runner, and that took time. Imagine that a runner from ancient Greece were just arriving here today, breathless, carrying the news that the Persians had been defeated at Marathon, and we were the first to hear it from his lips. If the distance were not 26.2 miles but billions of light-years, and if Greeks, like light, could live forever, this would describe the universe. Messengers in every form of radiation have been running for billions of years toward Earth and are arriving at our telescopes at every moment from every direction with news of their ancient eras—and our generation is the first to have both the technology and the theoretical subtlety to see and understand the messages. The cosmological revolution of today is the achievement of a few thousand human beings around the world, collaborating and competing to decode all the messengers' languages, understand the news, and pull the whole story together.

Just as a book can speak to your heart even though it was never addressed to you, those messengers are speaking to us intelligent beings and telling us, who are made of time, what that can mean.

Imagine that you have suddenly lost your memory. You look at yourself in a mirror, but there is only this moment. You are unaware of any past, even the very moment before this one. You are solid, your cells are real, your heart is pumping, but you have no idea how you got to the spot on which you stand.

Who are you? There is no answer.

You are not your family background, your personal history, the work you've done, your hopes for the future. These don't exist. You have no family, no associations. You're like a computer with hormones. You are listening to the latest music, buying the latest improved products, and believing the latest media interpretation of the world outside your room. This is all you know.

Now look in the mirror and see beyond the momentary you of today back to the you of years ago, to the child you once were, then the infant. Send your consciousness backward through time at lightning speed, down past your parents and grandparents, past the countless generations before them, through your ancestors roaming from continent to continent, through your primate ancestors, down through all the animals that preceded them, back through the earliest life, into a single cell, down into the complex chemicals that made it possible, down into the molten planet and the forming solar system, the birth of your carbon and oxygen and iron in exploding stars far across the Galaxy, back through the universal expansion to the creation of your elementary particles in the Big Bang. This is not fantasy. This is science: you are all this. Who you are is the sum total of your history. How far back you take that history—how much of your own identity you claim—is up to you. No one has had this choice before. It is only in the twenty-first century that we have learned enough to conceptualize the full scale of our genealogy.

To become a cosmic society we need to understand what we

humans are and how we fit into the evolving universe, both in our physical nature and in our potential significance. But more than that, the universe has to start mattering to us, as the ancient pictures of the universe mattered to our ancestors, or we will never get any benefit from the most astonishing knowledge that humans may ever have unearthed. We are stardust plus time. It took billions of years for galaxies to form; for stars to form inside the galaxies; for stardust to build up through many generations of stars and become common enough so that rocky planets could evolve; and then on at least one of those rocky planets, for life to evolve to the complexity of beings like us, who can reflect on it all and marvel. We are connected to the universe in our bones, our history, our atoms, and our minds. Scientific skepticism doesn't require that we maintain a feeling of detachment from the universe! Our connection is as real as anything can be and the strongest foundation for a coherent worldview in our time.

Chapter 5
This Cosmically Pivotal Moment

To see adequately into the future, we have to expand our view of the past. There is a kind of symmetry between past and future in everyone's consciousness: how far we can imagine the future is limited by how well we can conceive the past. Anyone can name a date ten thousand years in the future, but it's a meaningless number unless they have some sense of how much has changed on Earth over the past ten thousand years. By becoming aware of the multibillion year evolution of our universe, we begin to comprehend that the human future could be vaster than anyone ever imagined. Until recent cosmological discoveries created the concepts necessary to think on cosmic size scales, this understanding was not possible. But it's now becoming clear that we humans are living at an extraordinary moment in the history of the universe. If we don't soon begin not only to recognize this fact but to appreciate the future-shaping power of our moment and use it well, it could be lost forever.

There are millions of people in the United States today who believe that the world was created a few thousand years ago and that human beings have been on it the entire time except for the first five days. There are millions of people who believe that the end is near, that the Rapture is coming. But the evidence is now overwhelming that we humans were not here at the beginning and that today is nowhere near the end. We are living at the midpoint of time on multiple timescales.

For millennia it was assumed, as Genesis implies, that creation was about six thousand years ago. Then in the nineteenth century the science of geology took off, and when scientists tried to understand how mountains and river deltas formed, they were shocked to discover that some formations had to have taken hundreds of millions of years! This was a dizzying mental leap to "deep time"—time on the size scales of geologic change. Throughout the nineteenth century as excavation for canals and railroads went on in Europe and the United States, dinosaur bones and other fossils were uncovered. As scientists revealed this, the realization dawned on people that countless entire species had become extinct long ago. By the mid-twentieth century, scientists using radioactivity determined that Earth and the solar system itself are far older still—4.6 *billion* years old. Compared to this vast age, we humans just split from the other apes yesterday.

People carelessly use the word *infinite* to describe both space and time, but without a cosmological sense of time, *infinite* just stands in for *et cetera* or, as the novelist D. H. Lawrence put it, "a dreary on and on."[1] Imagining time as an endless "on and on" misses all the extraordinary things that can happen only on vast timescales. The ability to think on the timescales of the cosmos is new to our species; politicians certainly aren't using it yet, and in our daily lives it rarely seems essential to think much further into the future than next year, or when the kids go to college, or retirement, or the next generation. But this has to change. Our ideas of time have changed drastically in the past, and they can again.

We are living at the midpoint of time on the size scales of the cosmos, the solar system, Earth, and humanity—all at once—as we will explain below. The midpoints for the cosmos, solar system, and Earth tell us where we are, but being at a pivotal moment for humanity tells us what we need to *do*.

How likely is it that today really is a cosmically pivotal moment? The probability seems practically zero that in a 13.7-billion-year-old universe anything should be unique about the vanishingly tiny window called twenty-first-century Earth, so claiming that today is cosmically pivotal sounds like wishful nonsense. But don't look at the probability: look at the evidence.

We are living at the midpoint of time in four different senses.

First, on the cosmic scale, *this is the peak moment in the past and future of the universe for astronomical observation.* There will never again be so many galaxies visible.[2] Astronomers used to assume that although the expansion of the universe prevents us from seeing galaxies beyond the cosmic horizon, as time goes on more and more galaxies would come into the expanding cosmic horizon and become visible. But now we know that the opposite is true: because the expansion of the universe has begun to accelerate, the most distant galaxies are now disappearing over the cosmic horizon. The visible universe will empty out. There will never again be as many visible galaxies as we see today. When the universe is twice its present age, most distant galaxies will have disappeared beyond the cosmic horizon.

Meanwhile, as the distant galaxies get farther away, the neighboring galaxies in our Local Group—which are all bound together by gravity—are mutually attracted and are moving closer together. Over the next five billion years or so, the Milky Way will merge with Andromeda, forming a new galaxy that we can call Milky Andromeda.

▣ The simulation in figure 49 shows two massive spiral galaxies like the Milky Way and Andromeda merging. The galaxies approach. On their first close pass, the two galaxies sideswipe each other, throwing out long tails of stars and gas. They then move apart before coming back together again. First their centers merge, then the stars at greater distances from the center are tossed into random orbits, and eventually the merged spiral galaxies form an elliptical galaxy.

Eventually Milky Andromeda will be all that is visible to anyone on any planet inside, and it will be impossible forevermore to observe any other galaxies. The stupendously rich sky in the Hubble Ultra Deep Field, dense with galaxies, will be known to our distant descendants only historically through the records we leave. Those distant descendants' own deepest photos of space will show almost nothing. If we humans had not evolved to these abilities while the galaxies are still visible, it is possible that no intelligent beings in the distant future would ever be able to figure out how the universe operates. The astronomical observations and understandings that we pass on will be an irreplaceable part of the human heritage.

It took billions of years of cosmic evolution to build up the stardust required for the existence of rocky planets like Earth, and billions more years of biological evolution for creatures to evolve with the technological ability to study the distant galaxies. But we have appeared just as those distant galaxies are starting to disappear. It is in this sense that

Fig. 49. Two galaxies merging (time sequence is upper left, upper right, lower left, lower right)

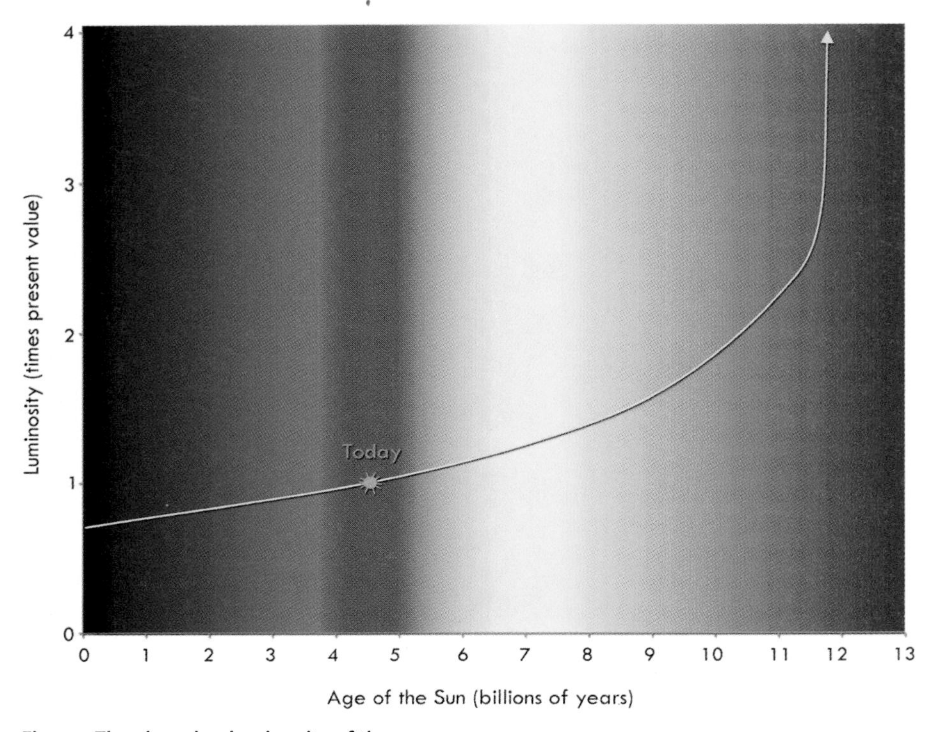

Fig. 50. The changing luminosity of the sun

this is the peak moment in the history of the universe for astronomical observation.

Second, this is close to the midpoint of our solar system, because the sun is now about 4.6 billion years old, and stars of its type follow a predictable life path that lasts about 10 billion years (fig. 50). The sun has 5 or 6 billion years to go, slowly heating up, before it puffs up into a red giant star and swallows the inner planets of Mercury and Venus, possibly burning Earth to a crisp.

Third, today is the midpoint of the existence of complex life on Earth. We humans have evolved in the middle of the approximately one-billion-year lush and habitable period for Earth, represented by the green zone on figure 50, when the planet has an oxygen-rich atmosphere and lots of water. This period—the age of animals—began only about half a billion years ago, after microorganisms had increased the oxygen content of the atmosphere to nearly its present level. But in another half a billion years, this habitable period will end as the sun's increasing heat output evaporates the oceans. The water vapor will rise to the top of the atmosphere, where the sun's ultraviolet light will dissociate it into oxygen and hydrogen, and most of the hydrogen will be permanently lost into space. Earth will then become a desert planet like the one described in Frank Herbert's science fiction masterpiece *Dune*—an extreme portrayal of the scarcity of water.

We modern humans have appeared at the midpoint of this great process: far enough into it to have a spectacularly evolved planetary ecosystem for a home plus all the powers of our memories, intellects, creativity, and passions, yet far enough from the end to allow extraordinary possibilities for the evolution of our descendants. The sun will provide Earth with a perfectly livable amount of heat and light for at least several *hundred million* generations—an almost unimaginably long time.

We might even be able to extend this timeline. As the sun heats up, our descendants will have millions of generations to prepare to survive. They may move to another suitable planetary system, but more interestingly, they could gradually move Earth farther from the sun and thus postpone catastrophe for billions of years.[3] Astronomers have already calculated how this could be done by changing the orbits of large comets so that they borrow energy from Jupiter and transfer it to Earth. Every hundred thousand years or so our distant descendants would need to engineer another close pass-by of a comet. Of course, they'd have to do it carefully, since if a comet accidentally hit Earth, they would go the way of the dinosaurs.

In these three ways, our place at the midpoint of time is something we humans did not plan, do not control, and had to discover, but the place is nevertheless real and gives us a big perspective we could not have had without this understanding. The fourth way we are central in time, however, is one that we also did not plan but can control and the one that is by far the most pressing.

The fourth way is that today is a pivotal moment in the evolution of our species. At the very moment that we are discovering our place in the cosmos, we are reaching the end of a period of explosive worldwide growth in both the human population and the physical impact of each one of us on the planet. This period of explosive growth has gone on longer than the lifetime of anyone now living, and therefore it seems normal, even inevitable. But from a larger perspective it is not normal at all and cannot last.

In 1800 there were about a billion people on Earth. In the past two centuries the population has increased by a factor of six, or six times. In the twentieth century alone the population doubled and then doubled again.

Let's look at a graph tracing the growth of the human population over the past two thousand years (fig. 51). Exponential growth always looks more or less like this curve: it rises slowly, then shoots up sharply like a bent elbow. Growth of something is "exponential" whenever the rate of growth is proportional to the amount of whatever is growing—in other words, the more there is, the faster the rate at which it grows. In biology, a species can get into runaway reproduction and grow ex-

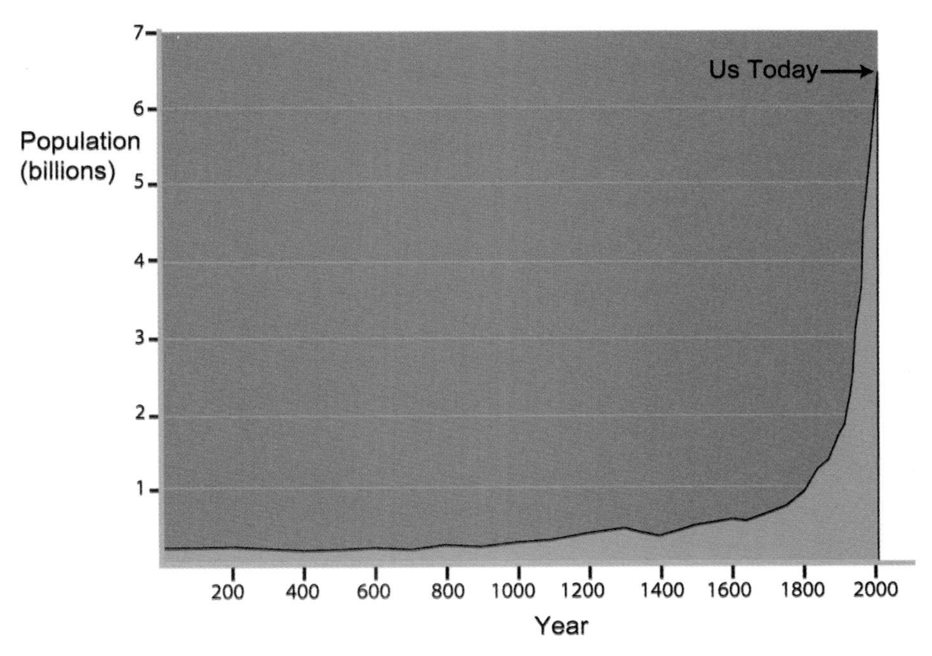

Fig. 51. Human population growth

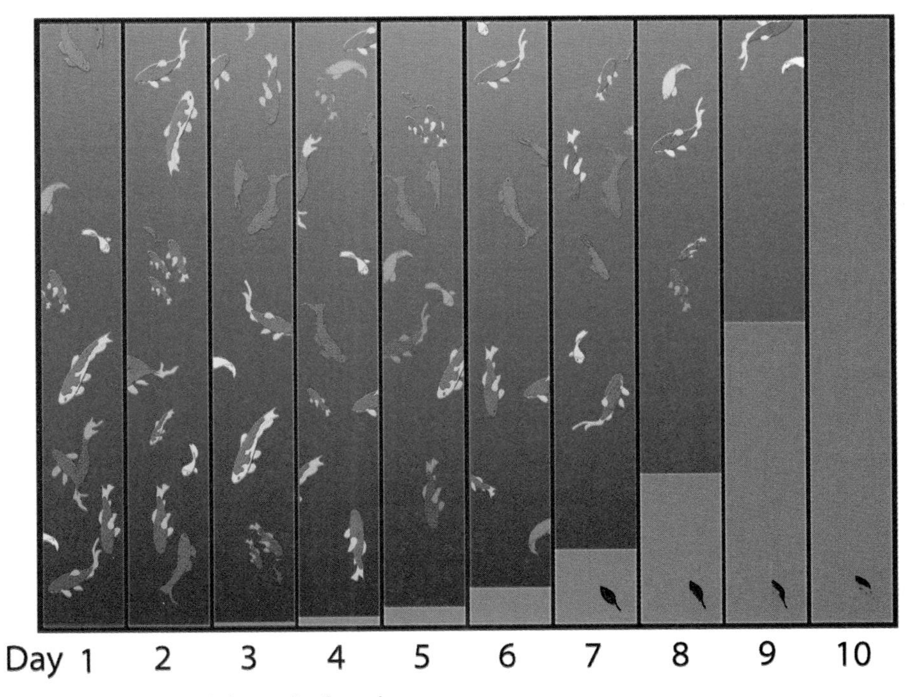

Fig. 52. The exponential growth of pond scum

ponentially, but if it then overconsumes the resources of its ecological niche, there is an abrupt die-off. Take, for example, a hypothetical bloom of pond scum that doubles each day. It starts slowly, but speeds up. Until the last couple of days, the pond looks nice and the fish are happy, but on the last day the scum chokes the whole pond and everything dies (fig. 52). It looks a lot like the graph of the human population over the last millennium.

As we write this book, the world population is approaching seven billion. Population experts agree that Earth cannot support another doubling of the human population. We will hit a limit before that. Hitting a limit is inevitable not only for the human population but probably even sooner for the exponential growth of natural resource use *by each person*. While population was increasing six times, carbon dioxide emissions increased twenty times, energy use thirty times, world gross domestic product a hundred times, and mobility per person a thousand times! If all the people in the world were to consume like Americans, which many aspire to do, it would take the resources of four Planet Earths. A typical person in the United States uses his or her weight in materials, fuel, and food *every day* (fig. 53). The United States and a few other countries have been dumping far more than our share of carbon dioxide into the atmosphere and oceans (fig. 54).

Clearly, we are beginning to hit material limits. Increases in greenhouse gases are now causing worldwide climate changes, the effects of which we are seeing in the form of record-breaking heat waves and storms and the melting of the polar icecaps. We are running out of fresh water and topsoil worldwide. We have destroyed more than half of the earth's forests and wetlands, and we are appropriating for our own consumption a large and increasing fraction of the biological pro-

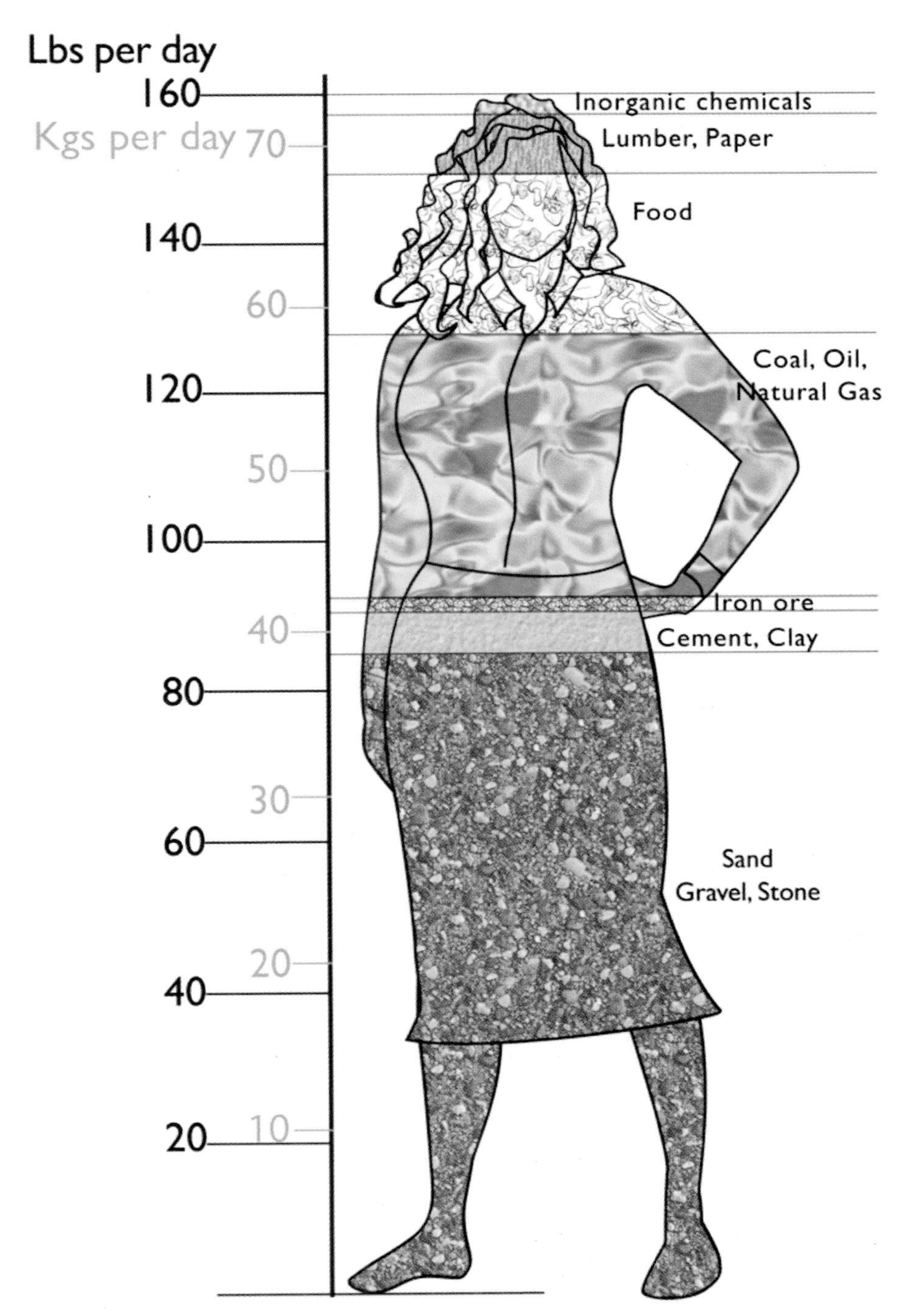

Fig. 53. The daily consumption of resources per person in the United States

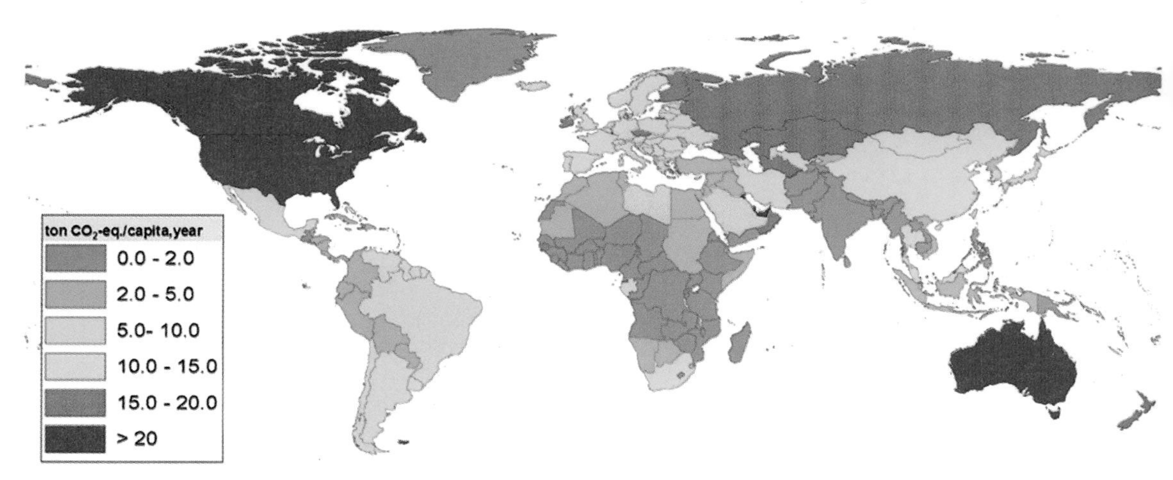

Fig. 54. World emissions of greenhouse gases per capita in 2005

ductivity of the entire earth. Our actions are killing not just individual organisms but wiping out entire species at the greatest rate since the extinction of the dinosaurs and many other species after the meteor impact sixty-five million years ago.

What most people do not understand, because it is counterintuitive, is how little time is left once an exponential trend becomes noticeable at all. The pond scum doesn't seem to be a danger to the pond until the next-to-the-last day. This is why we need to figure out *quickly* how to transition out of the current period of worldwide human inflationary growth as gently and justly as possible. Cosmology can help—by providing a model for this seemingly insurmountable task. The model fits because this pivotal moment for humanity is mirroring the most important pivot point in history: the beginning of our universe.

Our narrative is going to step backward here to explain what may have occurred in the instant leading to the Big Bang. Then we'll show the way humanity today is mirroring that instant and how we might be able to use this knowledge to transition out of this dangerous period in a way demonstrated by the universe to work.

According to the theory of Cosmic Inflation, just before the Big Bang (or at the very beginning of the Big Bang, depending on how you choose to look at it) there was a very brief period of about 10^{-32} seconds during which the universe expanded *exponentially;* in other words, in each successive unit of time it doubled in size, again and again. Then this exponential growth ended abruptly in what we call the Big Bang, after which the universe continued to expand, but far more slowly.

Cosmic Inflation is the only theory known that explains how the Big Bang could have gotten started—how the right initial conditions could have existed for the Big Bang to have happened the way it did. The theory predicts exactly the small differences from place to place that could grow with cold dark matter into the galaxy distribution that astronomers actually observe throughout the visible universe: the great chains, clusters, and superclusters of galaxies that lie along the filaments in the cosmic web. These small differences arose from quantum effects that occurred during cosmic inflation.[4]

The theory of Cosmic Inflation makes six predictions, and as of this writing five have been tested and found to agree with observations. The theory also appears to be compatible with modern particle physics theories, so it is definitely to be taken very seriously.

The shape of the curve representing cosmic inflation looks like the curve of human population or of pond scum—the only difference is the time between doublings, which for cosmic inflation was not years or days but an almost inconceivably tiny fraction of a second. If the theory is right, in the 10^{-32} seconds before the Big Bang the universe

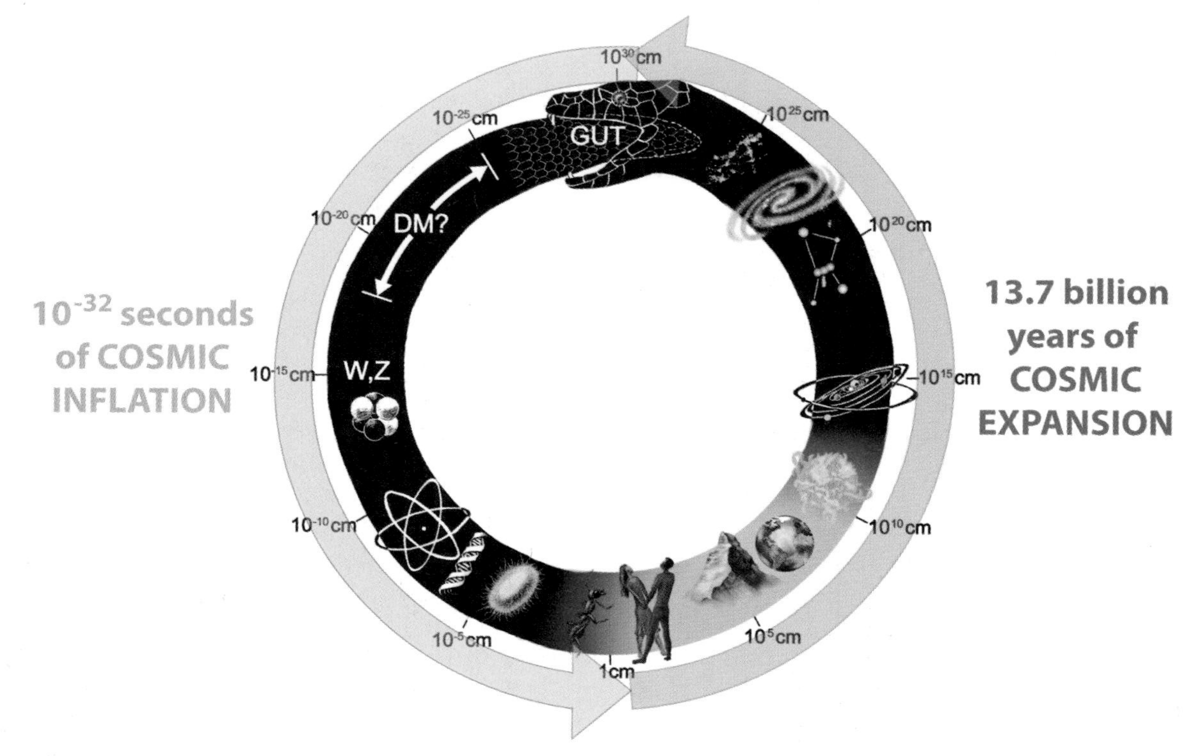

Fig. 55. Cosmic inflation and cosmic expansion

expanded just as much, in powers of ten, as it has expanded in the 13.7 billion years since! We can use the Cosmic Uroboros to illustrate that the size that the presently visible universe had reached by the end of cosmic inflation was fully halfway, logarithmically, to the size it is today (fig. 55).

The way that the universe transitioned from its exponentially explosive growth during cosmic inflation to the slow expansion that let it go on for billions of years could model for us the transition from rampant growth to sustainability that we humans must make. Countless cultures going back at least to ancient Egypt and Sumer used the cosmos as they understood it as the model for their lives. Now that we understand incomparably more about how the universe actually works, it is even more important—and valuable—to do this. The death of the pond is one model of how exponential growth can end; the universe gives us a very different model.

The universe's inflationary period ended abruptly with a Big Bang—but this was good! It was only *after* cosmic inflation ended and cosmic expansion became relatively slow that the universe entered its most creative and long-lived phase. The fundamental character of our universe has been to grow in complexity, but such growth takes a very long time. This could be the model for our future because long, slow growth is what we will need to solidify a sustainable civilization. But how did this transformation occur in the universe? We have to look a little more deeply into what was happening *during* cosmic inflation.

Quantum effects that were randomly occurring during the period of cosmic inflation also got expanded exponentially. When the exponential growth hit a limit and stopped, those quantum effects were permanently embedded in space-time in a unique pattern of wrinkles, like the lines on your hand. The pattern has continued to expand as the space-time in which it is embedded has expanded. The expanding wrinkles gravitationally attract dark matter, which has slowly been concentrating along them ever since, creating the expanding cosmic web. The pattern of wrinkles has been the blueprint for our universe. It was created during the brief inflationary moment, and the entire universe has been building itself on it. The effects of the fraction of a second of cosmic inflation will reverberate forever. That reverberation in some sense *is* our universe.

How can we use this as a model? Our own inflationary growth must end, but it doesn't have to be catastrophic. Afterward, if all goes well, it is still possible to grow, but only very slowly. The universe has shown that exponential growth transformed to slow growth can last for billions of years. But there is also a warning in the model: when cosmic inflation hit its limit, countless random events that were happening—quantum fluctuations—froze into permanent wrinkles in the new space-time. Amid partisan mudslinging in Washington and a "you first" attitude frustrating progress on the international level, it's tempting to discount the politics of our day. That, however, would be an irreversible mistake, because what the warning of the model means in practice is that countless political and social decisions being made on all size scales during these final years of human inflationary growth may end up getting frozen into the future of our species and our planet. Nothing could be less useful than to think that politics doesn't matter. Today's actions—and failures to act—may reverberate into the distant future far out of proportion to the thought going into them.

By the time our species' runaway growth becomes obvious enough so that everybody recognizes it and the detractors have shut up, it will be too late. Thankfully, we humans now know enough to be able to understand such dangers. Are we smarter than pond scum (fig. 56)?

Denial is easy. Today's humans may keep thinking small, keep bickering, refuse to face reality, and over the next few decades permit the destruction of the necessary conditions for decent human survival. The vast majority of Earth's species have gone extinct. There is no special dispensation for our branch of the primates. But this is no excuse for giving up on ourselves, because the very fact that we can under-

Fig. 56. Clay Bennett: "Now Playing—A Reassuring Lie"

stand this and discuss the options proves that we have the huge advantage of foresight. We have the power to tip the growth curve—and a small tipping of the curve today will have dramatic effects over time. (We discuss this in chapter 6.)

Many fine people genuinely trying to save our world assume that the only solution is for growth of population and of resource use to stop and then decrease. But if we take the universe as our model, we notice that at the end of cosmic inflation the universe in fact did not stop dead, like a truck hitting a brick wall. The exponential *rate* of growth stopped, but not growth itself. The universe slammed on the brakes, slowed to a crawl, and kept going for billions of years with no wall in sight. Inflationary growth in the human world that is transformed to slow but steady expansion can go on for billions of years.[5]

Another false assumption: for many people the end of growth is immediately interpreted economically as meaning the death of progress and freedom, but biologically the end of growth looks very different. We reach a certain size in adulthood. People whose bodies keep growing nonstop are suffering from the disease of gigantism; they become weak and die. The same may be true of an economy. This is the end of humanity's adolescence; it's a coming of age, and from here on the growth of complexity in human civilization shouldn't be physical any longer but intellectual, emotional, artistic, relational, and spiritual.

If we take the universe as our model, we should plan for and seek a *stable* period in resource use, which can happen only with renewable resources. The universe, of course, made its shift naturally. For it, injustice, suffering, addiction, and fatalism did not have to be overcome because they didn't exist, but for us they do. Nevertheless, we have the knowledge and internal resources to overcome them. Only resource-heavy activities have to slow down. Our drive for meaning, spiritual connection, personal and artistic expression, and cultural growth can be unlimited. These abstract treasures are often adequately appreciated only after they are lost, but if we valued them above consumer goods, then we would have a new paradigm for human progress. For our universe the most creative period, which brought forth galaxies, stars, atoms, planets, and life, came *after* inflation ended, and this could also be true for humanity. *A stable period can last as long as human creativity stays ahead of our physical impact on the earth.*

The goal should be sustainable prosperity, which is perfectly defined by the Zen saying, "Enough is a feast." The key to sustainable prosperity is to find the growth rate that allows human ingenuity to stay just ahead of resource use by foreseeing and minimizing consequences. People sometimes assume that to maintain a stable environment, innovation must be suppressed. But the opposite is the truth. Nonstop creativity will be essential to maintain long-term stability. Those of us who are alive today will largely determine whether this possibility will exist for our children, grandchildren, and their descendants—or not. Either way is a huge choice that will shape the character, for better or worse, not only of the long-term future but also of each one of us.

We humans might be the first. There may be microbial life on many other planets in this Galaxy, but it took a series of outrageously improbable events on Earth, plus multiple cosmic catastrophes to earlier species such as the dinosaurs, before humans could evolve. Earth is our only example of the evolution of life. If those improbable events were essential to make intelligent life, then our level of intelligence (and higher) may be extremely rare. Everyone is interested in discovering intelligent aliens out there, but suppose we are the only intelligence of our kind. Will there be consciousness and meaning in the future universe? Or will there just be sound and fury signifying nothing?

This is what it means to be living at a cosmically pivotal moment.

Many earlier cultures drew strength from the belief that they mattered to the universe, although this belief was based only on their mythology. Because today is the pivotal moment ending human inflation on Earth, we and our children may be the most significant generations of humans that have yet lived. If so, we matter to the universe, and our importance is based in science. If we wake up to the reality of our universe and our predicament on Earth; if we accept well-supported facts without letting ideologies suppress or distort them; if we become willing to expand our interpretations of our religious traditions to allow and value new knowledge, making those religions worthy of their time; and if we try our best to integrate the new universe into our thinking until it infiltrates our imaginations and our art, then our culture will have a new kind of Enlightenment. We will become a cosmic society.

Chapter 6
Bringing the Universe Down to Earth

There is a Native American concept that a person's responsibility extends "to the seventh generation." This is a wonderful impulse, but the phrase is wrong for today, not because seven is too many or too few but because it implies that every generation has the same level of responsibility. We who are alive today have a far greater responsibility than earlier, less knowledgeable generations or later, less pivotal generations. We happen to be the ones living at the end of human inflation. We need to choose a way to think about our responsibility that's appropriate for our time.

There are people who claim that human extinction might not be such a bad thing—that Earth would be better off without humans because we're wrecking it. But from the point of view of *the universe as a whole,* intelligent life may be the rarest of occurrences and the most in need of protection by those who can understand this. We—all intelligent, self-aware creatures that may exist in any galaxy—are the universe's only means of reflecting on and understanding itself. Together we are the self-consciousness of the universe. The entire universe is meaningless without us. This is not to say that the universe wouldn't exist without intelligent beings. Something would exist, but it wouldn't be a universe, because a universe is an idea, and there would be no ideas. Earth's problem is not the presence of intelligence: it's that perhaps we're not intelligent *enough* to take a cosmic perspective—yet. But the answer is to give ourselves time. Give our species time. And by definition the only way to do this is to live sustainably.

Not only are we humans at a pivotal point for our species, as we have discussed, but the choice of direction we collectively make could reverberate far beyond our species and even Earth. If we choose wrong, it could impoverish the consciousness of the universe, even destroy it. Relying on aliens to preserve the meaning of the universe is, at this stage of our knowledge, the same as delegating the job to possibly nonexistent angels. Our suicide could be the suicide of the universe if there are no other alien meaning-makers out there—and there might not be. We don't know. We few billion humans alive today represent through no democratic choice whatsoever the millions of generations of our ancestors and millions more generations of our potential descendants, with all their hopes and dreams and creations. We randomly-alive-today people actually have the power to end this evolutionary miracle, or not. Is suicide really the character we want to brand on the universe? Without human beings, as far as anyone knows, the universe will be silenced forever. No meaning, no beauty, no awe, no consciousness, no "laws" of physics. Is any quarrel or pile of possessions worth this?

The simple truth is that the future of our species may depend on what we do now to protect the conditions of life. Everyday human actions are already having effects on planetary and even cosmic timescales, yet most people are still obliviously living in an obsolete cosmology where these effects are invisible; they are living in an illusion of a universe that does not exist. From their perspective it's impossible to conceptualize and get a handle on what we humans have to do to preserve ourselves and all our potential futures here in the real universe. Now let's be clear: Earth itself is in no need of saving. It's been through far worse—frozen, steamed, bombarded with meteors—and it will go on cycling for billions of years. What needs saving is the combination of conditions under which we humans evolved and thrive.

So, what can we really do? How can we make the choice to change direction and priorities? This chapter is about digging in and applying cosmic perspective to immediate problems.

It may seem strange that there should be a practical connection between the vastly different timescales of cosmology and our present environmental challenges, but not only is there a connection—it is crucial that people realize this very soon. By understanding how humanity fits into the timescales of the universe, we begin to grasp what is truly at stake for our planet and for our descendants in the political and ecological decisions being made today. The stakes are far greater than most members of the public appear to appreciate. But we have the unexpected opportunity to exploit some of cosmology's concepts to help ourselves conceive, comprehend, and begin to care and act on those vast scales.

In his election night presidential victory speech, Barack Obama told of a 106-year-old voter he'd met named Ann Nixon Cooper and all the momentous changes she had seen in her lifetime. He laid down a challenge that first night to think one hundred years ahead, specifically to a time when his two young daughters, Malia and Sasha, may still be living. "If our children should live to see the next century," he said, "if my daughters should be so lucky to live as long as Ann Nixon Cooper, what change will they see? What progress will we have made? This is our chance to answer that call." Thinking one hundred years ahead is virtually never done in Washington, and yet, as Obama pointed out, many people who will be affected then by the consequences of today's decisions are already here—they are citizens, and they should count. Instead of thinking of an abstract century, which is basically an arbitrary number, what if we think about the actual lifetimes of our very young children. Suppose we let "Malia and Sasha" symbolize today's elementary school generation, and we call the time one hundred years after the speech, about 2108, the "Malia-Sasha Horizon." Let's look at what one trend is telling us right now about the changes that many in this rising generation will actually have to cope with.

The blue line in figure 57 shows that the total concentration of carbon dioxide in the atmosphere has fluctuated within a very narrow range for the last two millennia. Analysis of air bubbles in Antarctic ice shows that the range for at least the past eight hundred thousand years was between a high of 300 parts per million (ppm) and a low of 170 ppm during the ice ages. The red curve on the graph reflects the amount humans have contributed, which has been doubling roughly every thirty years since about 1800. The very rapid increase in the total carbon dioxide in the atmosphere is due to this human contribution. What's shocking is how high that blue line would go at this rate by the time it reaches the Malia-Sasha Horizon (the green line).

Two alternative futures for the amount of carbon emitted between now and the Malia-Sasha Horizon are projected in figure 58. These are two scenarios from the report of the International Panel on Cli-

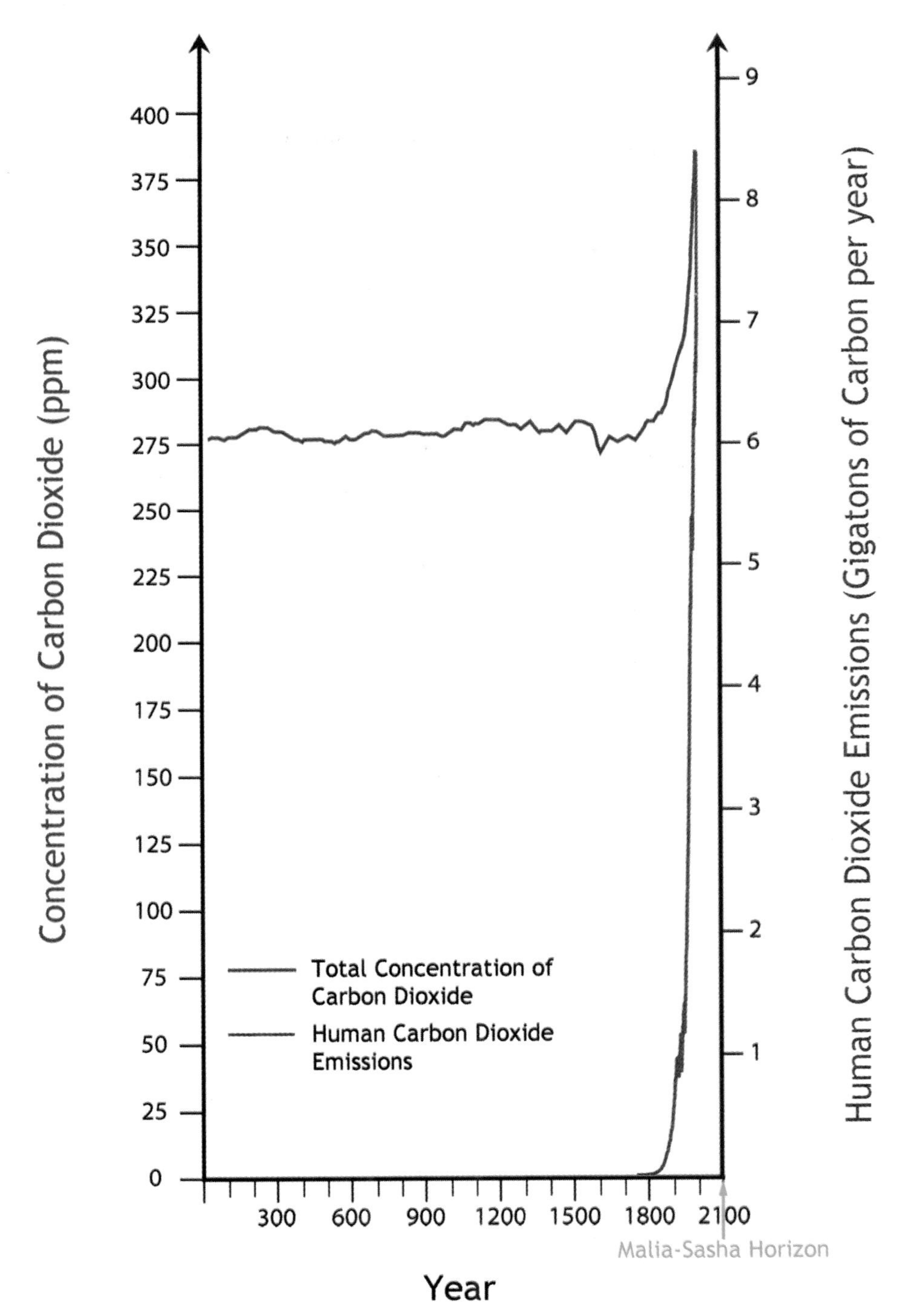

Fig. 57. The concentration of carbon dioxide in the atmosphere, with the exponentially growing human contribution

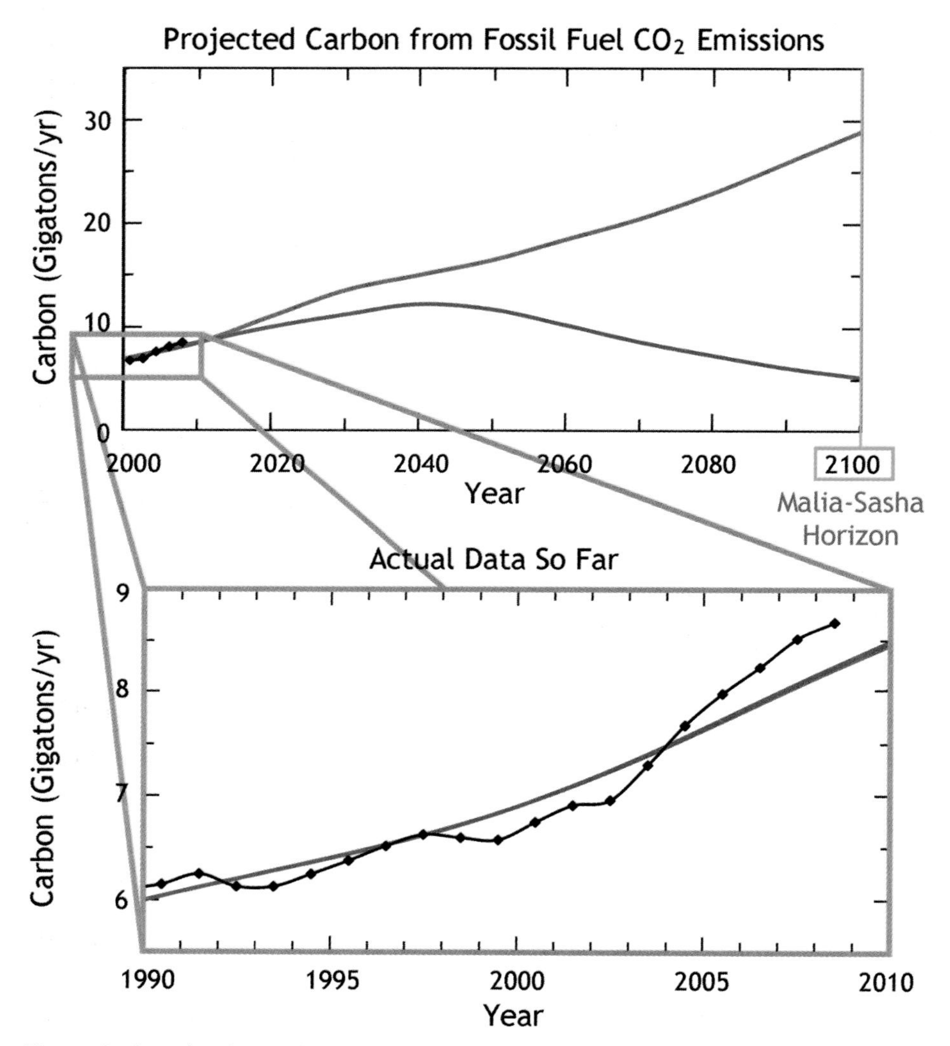

Fig. 58. Projected carbon emissions through 2100 and actual data so far

mate Change, which shared the 2007 Nobel Peace Prize with former Vice President Al Gore. The red line represents what will happen in a business-as-usual scenario, and the blue line represents an optimistic best-case scenario with a major worldwide commitment to cut carbon dioxide emissions. Below the graph of projections is a blow-up of the time period that has just passed; those data (shown as a black line) have been measured, not projected.

In the top graph, business as usual is projected to lead to huge carbon emissions and a disastrous change in climate. But even the blue-line best-case optimistic scenario will lead to serious climate change. Knowing this, how are we doing? Are we reducing our emissions, or is it business as usual? Let's look at the actual data points on the bottom graph: we're emitting far *more* carbon than even the business-as-usual projection of a couple of years ago.

In figure 59 we see how temperatures will look across the United States at the Malia-Sasha Horizon under the two projected scenarios. These are frightening answers to Obama's questions, "What change will they see? What progress will we have made?"

Economists have taught us to discount the value of future goods, but people have carelessly, and self-interestedly, extrapolated that idea to justify discounting the very reality of the future and of the people who will be living then. We dismiss them as if they won't be real. If our political leaders, business leaders, cultural leaders, and voters simply began today to consider as *real* the impacts of their own decisions out to the Malia-Sasha Horizon, this would be a gigantic first step in expanding our practical consciousness to include cosmological time.

The first step is the most important. It sets the direction and overcomes inertia. Becoming aware of our enormous potential future

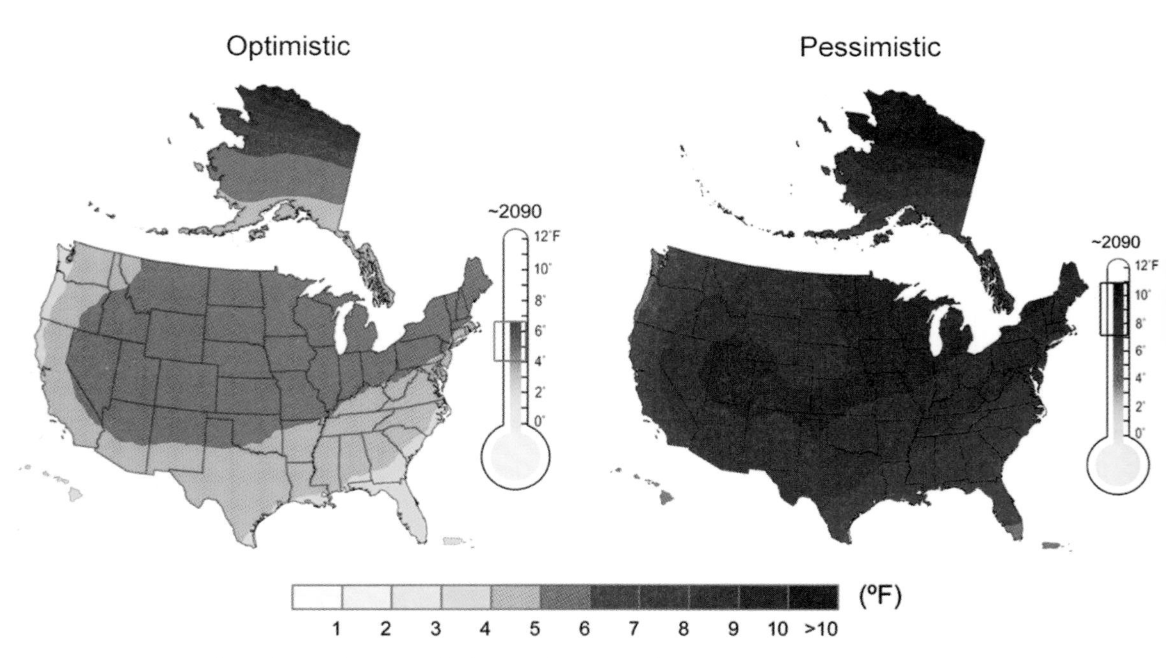

Fig. 59. Optimistic and pessimistic temperature scenarios

strengthens our spirits right now as well as our sense of commitment to the long-term future. But to do the best practical job we can to create a cosmic society, we don't have to think extremely far ahead. All we have to do is make relatively small but immediate changes to *alter the trends,* and if we persevere in those small changes, then after a few decades the results will be huge. If we have been rigorous in analyzing our situation and honest in our interpretations, those changes will take us in the right direction, and we may achieve sustainable prosperity.

This chapter provides two stories of practical actions that, if begun now, could avert enormous long-term problems. The first describes a way to proceed when the science on which a wise decision should be based is uncertain and controversial yet action cannot be postponed. The second describes a situation that is globally pressing and where there is virtually complete consensus among scientists—yet no one is acting, because they do not or will not recognize the dire long-term consequences of inaction.

Official Action Despite Scientific Uncertainty

In 1976 coauthor Nancy Ellen Abrams was a young lawyer at the Office of Technology Assessment, Congress's in-house science advisory office until Congress abolished it in 1995. The Office of Technology Assessment already had convincing evidence in the 1970s that the average temperature of Earth was warming, that cheap oil was running out, that our health care system was unsustainable. The job of Abrams's section was to analyze the country's long-term science and technology needs, but the longest anyone ever considered was about thirty years into the future. In Washington, where major decisions are often made with an eye to the current crisis or the next election, most people considered thirty years an absurdly luxurious and highly unrealistic timeline, and they didn't take it seriously. The result is that now those thirty years have passed, and every one of those problems has only gotten worse. Now time is running out—and because of the nature of exponential growth, it is running out faster and faster. It's clear our society must change course, but first we have to figure out how to do so, and that requires understanding the place of science in the big decisions of our time.

There is no question that we have to stop bingeing on carbon-based energy sources, but there are serious questions about how best to do that. One of the big questions, for example, is the potential role of nuclear power. In the 1970s and 1980s the dangers of nuclear power combined with its huge expense convinced many of us that nuclear power was not a viable or cost-effective long-term solution. It requires immense government financial aid, and even today, the United States still hasn't figured out how to dispose of nuclear waste in a way that will be safe for the tens of thousands of years it will remain radioactive. These are major negatives for nuclear power. But on the other hand, we have a potentially much bigger crisis on our hands if humanity keeps pouring carbon into the atmosphere, the climate becomes chaotic, ocean currents change, and around the world there are extreme droughts, hurricanes, floods, crop failures, wildfires, diseases spreading where they never appeared before, and countless climate refugees, both humans and wild animals, roaming the globe and fighting just for a place to live. So on second thought, if nuclear power could seriously help avoid this fate, we have to reconsider it with an open mind and ask, could it be made safe and economically competitive? How so? And which version would be best?

But whom can we trust to answer these questions? Not the energy companies, whose goal is their bottom line. Not the relevant government agencies, besieged by lobbyists and the media, and juggling economic, political, and social pressures; they rarely have the ability to figure out the scientifically most responsible answers.

Abrams invented a method to find those answers while she was working at the Office of Technology Assessment, and she developed it in an article coauthored with R. Stephen Berry of the University of

Chicago. The method is called Scientific Mediation.[1] Its purpose is to separate out the underlying scientific questions embedded in a controversial political issue like energy policy, so that they are given the importance they deserve, and so that not only those making the ultimate political decision but also the other stakeholders and the public can see what is known and what is not known, in order to make possible a more informed and socially responsible decision. Scientific Mediation is not intended to decide any political questions or ask any moral ones, only to clarify the underlying science. Scientists and other expert advisers to government are human beings like everyone else. They can mix personal biases into their supposedly expert opinions, sometimes without realizing that they are doing so. But during a Scientific Mediation those biases tend to surface, and it becomes clear to what extent a policy recommendation is truly based on science. When the report is made public, anyone can understand the technical part of the dispute and what is at stake—far better, perhaps, than the scientists themselves did in the beginning.

Scientific Mediation has never been attempted in the United States, but the Swedish government had great success with it in determining whether the Swedish utilities' plan for nuclear waste disposal was "adequate." (Sweden was compelled by law to certify that it had an adequate plan for nuclear waste disposal before it could open three completed nuclear power plants.) The country commissioned over forty simultaneous studies of their proposed nuclear waste disposal plan, by atomic energy commissions and expert university research groups around the world. For their own in-house study, the Swedish Royal Energy Commission decided to use Scientific Mediation and hired Abrams to play the role of mediator. The Scientific Mediation, involving just three people plus an advisory group, discovered problems with the plan that none of the other much larger studies found. This result was so illuminating that Scientific Mediation became for some years standard procedure in the very forward-thinking Swedish Ministry of Industry. It was a brave choice, because Scientific Mediation is dangerous for an agency that has decided in advance what it wants to do and is merely looking for the blessing of experts—because, like the scientific method, this procedure coaxes out truths that may be inconvenient in the short term but are essential to know for the longer term.

To conduct a Scientific Mediation a government agency needs one well-respected expert scientist to represent each side of a science policy controversy and a mediator to facilitate the unusual protocol. Each scientist believes that the existing but admittedly inadequate data favor his or her interpretation. With the help of the mediator the opposing scientists collaborate to write a joint report in ordinary, nontechnical language.

1. They list the main areas of agreement, narrowing the dispute.
2. They not only state the arguments for their own side but also make the other side's strongest case, until they satisfy their opponent that they understand his or her position.
3. They list their main points of disagreement.
4. They then must *agree on why they disagree* on each of those points.
5. They finish by outlining essential research that must be done to answer the question adequately.
6. They both sign the document.

If some government agency or private foundation sponsored a genuine Scientific Mediation on the question of whether nuclear power could be made safe and cost-effective—or whether clean coal is possible, or what really is the maximum safe level of carbon in the atmosphere, or any of the other big scientific questions underlying current political debates—not only lawmakers but we the people would come out of it with the best understanding possible at this time, and one that would be fair to all sides. That alone would be invaluable. What the country then decides to do about it we can battle out! That's what politics is for, and a cosmic society will still be a political society. But the public should demand that *all options being fought over are physically realistic and feasible.* A cosmic society needs a sharable picture of reality that is based on sound science and from which we can proceed to build our future.

Official Evasion Despite Scientific Consensus

In 1988 the two of us traveled to the former Soviet Union in a collaborative effort of American and Soviet scientists and international lawyers to end the practice of launching nuclear reactors into orbit to power satellites. We wanted to help prevent a Cold War arms race in space and also prevent reactors from eventually becoming radioactive space debris. The American leader was Joel R. Primack, and the Soviet leader was Roald Sagdeev, who was then the director of the Soviet Space Research Institute. The Soviet Union had already launched about forty reactors into orbit, and two had fallen, one over northern Canada in 1978 and the other in the ocean in 1983. The United States meanwhile was planning to launch much bigger reactors to power missile defense satellites, popularly known as Star Wars, and we felt that the Soviet Union might be willing to stop if the United States would not start. Our delegation actually persuaded the Soviet government to agree not to launch any new reactors for two years. By the end of that time the Soviet Union had ceased to exist, and no additional reactors have been launched. This was a success in preventing the worst kind of satellites, but it was only a small part of an effort that has barely begun—to protect Earth from hundreds of other potential disasters from space debris.

Our society has become dependent on satellites. Their role in communications, global positioning, monitoring weather and climate, and other research is by now enormous. If humanity's future development is not to be based on endless resource consumption, it will have to be based on improved cultural links. We have to protect the means of communication. We'll need satellites for a very long time—but we can't keep launching new ones indefinitely without bringing down the old ones and preventing the accumulation of space debris.

Many people heard about space debris for the first time on 10 February 2009, when an American "Iridium" communication satellite and a defunct Russian satellite, moving in different orbits at 27,000 kilometers per hour (17,000 miles per hour), crashed into each other and broke up into countless pieces. When that kind of thing happens, it's not like the science fiction movies. In the first *Star Wars* movie, for example, the audience sees a big explosion in space, and then a moment later, the pieces disappear and the view from the spaceship's cockpit is completely clear. What really happens when there's a collision in space is that pieces of all different sizes and shapes fly out. Some fall toward Earth and burn up in the atmosphere, but the rest go into erratic orbits at 27,000 kilometers per hour, which is ten times faster than a high-powered rifle bullet. As the number of satellites and pieces of satellites at a given altitude increases, pieces hit other pieces or satellites, smashing them into fragments, which will in turn hit more pieces, eventually starting a chain reaction. This process may already have begun.

About twenty thousand pieces larger than 10 centimeters (about 4 inches) in size are now tracked, but countless smaller ones are not tracked. No one knows exactly where they are, so there is no way to plan to avoid them. Figure 60 is a kind of snapshot of tracked pieces orbiting Earth at a given moment.

Space is our most fragile environment—it has the least ability to repair itself. Only Earth's atmosphere can remove satellites from orbit. Every eleven years, in a so-far unexplained cycle, the sun flares up. These flare-ups heat Earth's upper atmosphere and make it expand so that debris and satellites in low orbits are subjected to increased drag and start to come down. Most fall into the ocean, since Earth is mostly ocean. If they fall uncontrolled, there is some danger of their falling on land, but at least this is a means of cleaning space. The higher a satellite's original orbit, the less air there is to collide with, and the longer it takes for debris to reenter the atmosphere. There is no practical way to clean up space debris. Any bucket launched to catch debris would just become debris itself. Terrible as they are, millions of land mines left from earlier wars in Afghanistan and other countries can, with human will, eventually be removed, but debris in orbit higher than about 800 kilometers (about 500 miles) above Earth's surface will be up there for decades, above 1,000 kilometers (620 miles) for centuries, and above 1,500 kilometers (930 miles) effectively forever.

When you realize how long human civilization could continue and how much we need to do *now* to protect that future, you grasp the importance of our space environment. Once we understand that we are

Fig. 60. Space debris in Low Earth Orbit

at the midpoint of time, it becomes common sense—a new universe common sense—to insist that every satellite launched by anyone in the world, private or government, have a rocket on it that allows it to be brought down safely. The reason it's not done is that it would involve a small added expense that's not legally required. But humanity gets just one chance to save our space environment for all future generations, and now is it.

Accidental collisions of space debris, however, are not the worst threat to space: the greatest threat is war in space. The public debate over Star Wars has dragged on in the United States for a quarter of a century already, since the Reagan years, and it has focused on the immense costs, financial and political, and the unlikelihood that such a system could physically work. But the debate has almost completely missed a crucial point: even one war in space will create a battlefield that will last effectively *forever,* encasing our entire planet inside a shell of speeding metal shards. This will make space near the earth highly hazardous for peaceful as well as military purposes. No actual space war even has to be fought to create this catastrophe. Preparation for it is enough, because any country that felt threatened by some other country's weapons in space would only have to launch the equivalent of a truckload of gravel to destroy the sophisticated weaponry—but gravel would also destroy any civilian satellites at that altitude, including those we all depend on for weather information, global positioning systems, and communication.

These should be international space protection principles:

1. Do not introduce attack weapons into space.
2. Avoid fragmentation of satellites.
3. Prohibit explosions of any kind in space.

4. Require all satellites to safely reenter when their useful life is over.
5. Ban nuclear reactors in orbit.

Political alliances are always shifting, sometimes drastically. The deadliest enemies of the United States a mere sixty-five years ago, the Germans and Japanese, have been among our closest allies for decades already. It may seem that jazzy new weapons in space will give us a leg up over the enemy of the day, whoever it may be, but in making this choice we become our children's archenemy. Any temporary military advantage would pale before the overwhelming, eternal immorality of imprisoning Earth for thousands of years in a halo of bullets.

We might also consider the fact that our planet is cosmically exposed. Who knows what a halo of bullets would communicate to any intelligent beings that may be observing us? It certainly wouldn't speak well of our own intelligence.

No one can predict what the world will actually be like when the elementary school children of today are elderly, but we nevertheless have a moral obligation to take responsibility for the likely impacts of our own current decisions *at least* out to their horizon. And we have a corresponding intellectual obligation to base those projections on solid science, which will require new approaches like Scientific Mediation to get that level of science into the policy-making process not only of governments but of business, courts, public interest organizations, and all groups with influence on—and a stake in—the future.

From a cosmic perspective, we are living in the middle of the best period for life on Earth, which falls at the midpoint of Earth's lifetime and also the sun's, and which is also the peak moment in the entire existence of our universe for astronomical observation. Why all these midpoints are coinciding is not the point; perhaps it's just coincidence. What matters is that this knowledge gives us a way to grasp how our own far more pressing moment—the end of exponential growth, humanity's turning point, led by us whether we like it or not—is part of a cosmic endeavor with a larger meaning than most of us have ever considered. Putting this together, we are facing three challenges.

Our *first* challenge is to break through and *see* the new cosmos not just as a new idea in physics but as our shared mental homeland—a homeland where cosmological time is the only appropriate perspective on many issues, and global threats that may not spin out of control for another generation or two are nevertheless as *real* as a hurricane that will hit tonight.

Our *second* challenge is to use this new knowledge to develop a long-range, large-scale vision that can be widely understood and shared, irrespective of religion. The vision must be grounded in scientific understanding of both the universe and the idiosyncrasies of human consciousness, since our reality depends, and always will, on the interplay of both. A coherent, shared cosmology was for our ancestors a source of bonding power that enabled them to trust and cooperate in ever larger groups—and developing the ability to cooperate in increasingly larger groups made civilization possible. The emerging global civilization will need to cooperate in much larger numbers than humans ever have before. Such rapidly growing countries as China, India, and Brazil must be included in this consensus. Our scientific understanding of our human predicament may be the most important thing that we all have in common.

The *third* and ultimate challenge for all people is to seek to understand nature *in order to harmonize our behavior with nature,* not just to exploit nature technologically while generating heaps of garbage and unhappiness.

Will we rise to these challenges?

No one knows. The human future is undetermined. One thing, however, is absolutely certain: if enough of us commit ourselves to try, in light of the now known laws of the new universe, and if in doing so we build a de facto transnational community that values and supports this work and appreciates the riotous variety of contributions that create a unifying, visionary cosmology, we will dramatically increase the probability that the answer will be yes. The better we grasp the scope and meaning of the past, the better we become able to grasp the scope, meaning, and possibilities of the human future. This shared awareness can give us not only analytical tools but a bone-deep sense of urgency and hope that together can help us rise to what our pivotal time demands.

Chapter 7
A New Origin Story

The lack of a meaningful universe in our modern culture impoverishes every one of us, but the complications of daily life distract us, so few people ever stand back and notice that something essential to all human life is missing from ours: we have no believable, shared context for the problems we face together. We simply keep thinking locally, through the tired metaphors of our old political and economic systems and even older religions—while the effects of our collective actions radiate around the planet and out into the distant future beyond our current ability to conceptualize, comprehend, or care. The meaning or significance of anything a person does is never intrinsic to the action but exists only from the perspective of a larger context. Therefore, if we have no such context, we have no meaningful way of choosing. Despite interested parties yelling in both our ears why we should go one way rather than the other, we are often paralyzed. The easiest course is to go with those who promise what we assume we want. The consequences of this narrow thinking—the repeated choice of minor short-term good over major long-term good—are evident in public life, though their deepest cause is not. That is, not yet.

Whether you're a voter or the president of the United States, to understand what is happening on the *global* level, you need to think cosmically, not just globally. Cosmically is the larger context. And a prerequisite for thinking cosmically is having a meaningful cosmos. Earlier cultures showed us how to create one: it takes, perhaps before all other things, a believable story that explains how everything came about and how each of us is intimately connected to all that is. The story must explain the invisible forces in the universe. No one in an age of computers and other electronics based on quantum mechanics, and of global positioning systems based on general relativity, can seriously argue that there are no invisible forces. On the Cosmic Uroboros, almost all of reality is invisible. Then the story must convincingly explain our own existence here and now. And it must be equally true for everyone. This is so important a criterion that it deserves capital letters: Equally True for Everyone on Earth. For the first time such a story is possible.

The well-known mythologist Joseph Campbell, in his final book, *The Inner Reaches of Outer Space,* argued passionately that what the modern world needs more than anything else is a story that unifies. "The old gods are dead or dying," he wrote, "and people everywhere are searching, asking: what is the new mythology to be, the mythology of this unified earth as of one harmonious being?" We humans are emerging right now from an era of imaginative origin myths that has lasted many thousands of years. We are entering a new era in which our origin myth will be both inspired and verified by science. An assortment of people ranging from scriptural literalists to corporate polluters to post-modern philosophers denounce or dismiss science, but if those people ever take a modern medicine or fly on a jet, they're trusting their lives to science, and their actions are a more reliable sign of where their trust lies than their words. Some groups will always cling to old ideologies, but science is now a worldwide collaboration, and its discoveries are available to everyone. Science—not just modern cosmology but the entire scientific approach to reality—is the only possible foundation for a globally unifying story of ourselves. What we build on that foundation will involve far more than science—wisdom, daring, immense creativity, and good faith for starters—but the foundation must be solid or everything else will collapse.

Dedicated people around the world are working on solutions to various global problems, and they're experimenting with new ideas and technologies. But even if some group came up with a brilliant and complete blueprint for renewing the earth ecologically, politically, and culturally, leading to a sustainable, vibrant, and equitable worldwide civilization, it's highly unlikely that people today could agree on implementing it. A blueprint for saving the world, fantastically useful as that would be, would not on its own convince the earth's population that success is possible and therefore worth the cost today. What could do this, however, is a believable, desirable vision that includes us all. We humans are motivated not by ideas but by feelings, although ideas are the means through which we can use those feelings. We need to *feel in our bones* that something much bigger is going on than our petty quarrels and our obsession with getting and spending, and that the role we each play in this very big something is what really defines the meaning and purpose of our lives. This is what a cosmology has traditionally

Fig. 61. John Andrews, *Boston Tea Party—Destruction of the Tea in Boston Harbor, December 16, 1773*

done. Many people, religious and nonreligious, intuit this, but they can't agree on what the big picture is or what it demands of us because their individual views have no framework in which to contemplate the global size scale of our problems. But with a cosmic framework it becomes clear that we humans are playing a cosmic role—whether individuals believe it or not.

The importance of an origin story is easier to see on a smaller scale. Every country has its own origin story, usually including at least some truth, and it becomes the guiding myth of the people. For example, in the United States, every child learns the story of the Boston Tea Party (fig. 61), the ride of Paul Revere, the Declaration of Independence (fig. 62), the Revolutionary War, the Constitution, the Bill of Rights, the wisdom and foresight of the Founding Fathers. All these are

Fig. 62. John Trumbull, *The Declaration of Independence, July 4, 1776*

elements of the American mythology that binds our citizens together in a shared venture of upholding freedom and democracy, although none of us were there to witness any of these things. Even our name, the United States of America, tells a story. In the same way but on a far larger scale, to bind a global community together in the shared venture of preserving and protecting the conditions for our children and grandchildren to thrive on this blue planet, we humans need to see that the through line of our species back beyond our earliest ancestors and through the preceding cosmic events is *everyone's* origin story.

Earlier cultures developed metaphors that expressed their own mythic sense of origins: the World Serpent mated with the World Egg; Grandfather Fire created light, colors, and song; the spirit of God hovered over the Deep. We have to do as well for the universe that we now know exists and is ours. We have a new technical scientific understanding, but making it into a story and an accompanying mental movie is essential if we're going to have a mythic context that transcends cultural differences and is Equally True for Everyone on Earth.

The science-based origin story is about events on a different scale from any traditional creation story. It's the story of the new universe. It explains how we intelligent beings came to be part of this evolving cosmos and how deep and ancient an identity we share with each other and all life. It is mind-expanding and profound, transcends all local differences, and above all is supported by the evidence. But it feels unsettling at first—it doesn't have familiar characters or follow the implicit rules of storytelling or even the rules of existence on Earth. This can't be helped, because our human role goes beyond Earth. We are part of a phenomenally rare and cosmically important event: the emergence of intelligence and civilization in a universe that was once nothing but particles and energy.

"In the beginning" is the way origin stories around the world traditionally open. "In the beginning the high god rose out of the Primeval Waters," goes a creation story from ancient Egypt. "In the beginning there was no land. The Giver and the Watcher sat outside their sweat lodge," began a Native American version of creation. "In the beginning, God created the heavens and the earth," opens the book of Genesis. But just plain "in the beginning" doesn't work for a scientific story. "In the beginning" *must* be followed by "of" and then something definable. There is no beginning in itself. A beginning is a way *people* draw a line to help ourselves think about what may exist. We can say, "In the beginning of Earth" or "In the beginning of our universe," but we can't say "In the beginning," because that implies "of everything," and "everything" is not a clear notion until you can define it—which you can't if you don't have a convincing origin story. The first step is to be precise as to what our origin story is about: our universe, which, as we shall see, may not be everything.

Our New Origin Story—Stage One

In the beginning of the Big Bang was cosmic inflation. What would become our present visible universe was just a "sparkpoint"—a region so tiny it lay nearly at the tip of the serpent's tail on the Cosmic Uroboros. In a wild 10^{-32} seconds the sparkpoint inflated exponentially by at least thirty orders of magnitude to the size of a newborn baby; that is, it expanded across every size scale from the tip of the tail of the Cosmic Uroboros halfway around to the size of humans, as illustrated in figure 55. In that brief process it spawned all the quantum impulses that have shaped the cosmic web and will do so effectively forever.[1] Then suddenly, like a drop of water freezing to ice, the universe went through a phase transition called the Big Bang, and the rate of expansion dropped drastically from exponential to slow and steady. It has taken almost fourteen billion years to expand the universe by the same factor as happened in that first fraction of a second.

Space-time emerged from the Big Bang smoother than the smoothest lake, but not perfectly smooth. It had subtle wrinkles perhaps as tiny as elementary particles, wrinkles as big as the universe, and wrinkles of every size in between. This was good for us, because if the universe had been absolutely perfectly smooth in the beginning, then it would still be smooth now, only greatly expanded. There would be no concentrations of matter: no galaxies, no stars, no planets, no life. It has been the imperfections in the universe, the primeval wrinkles, that are the blueprint of the cosmic web of galaxies, galaxy clusters, and superclusters throughout the universe. But don't be confused by the metaphor: unlike an ordinary blueprint, the cosmic blueprint is not a sheet of paper separate from the construction; it is the foundation itself.

The new space-time was filled with a hot, dense, incredibly smooth, and nonturbulent expanding fog of particles, including quarks, electrons, neutrinos, and dark matter particles, all flying about at high speed, free and unbound. As the universe expanded and cooled, the particles swiftly revealed that they had distinctive natures—annihilating, interacting, fusing, gravitating. Light from the Big Bang was so intense that it hammered the electrically charged particles—electrons and protons although not dark matter—and yet the light could not get through them. Thus the baby universe was opaque. It took four hundred thousand years of expansion for it to cool enough so that the universe became transparent. Now the bottled-up light of the Big Bang streamed in all directions, carrying the image of the universe as it looked at that early moment, only four hundred thousand years after the beginning. That image is depicted in figure 63; this is the whole sky's light of the Big Bang.

As the young universe expanded, dark matter everywhere moved sluggishly toward the subtle wrinkles in space, pulling atoms along.

Fig. 63. Cosmic microwave background radiation

When regions containing dark matter became twice the average density around them, they stopped expanding and formed invisible halos, inside of which the visible galaxies would later form.

Gravity has been fighting expansion since the Big Bang itself. Expansion and contraction are the two counterbalancing forces of the universe. In those regions dense with dark matter, gravity has won and tamed that region of space. Galaxies and clusters of galaxies are gravitationally bound and will stay together forever, traveling as a unit in the great expansion while evolution goes on inside them.

But dark energy, which is created by space and was insignificant when space was tiny, has kept increasing as the universe expands, while the amount of dark matter stays the same. About five billion years ago dark matter—once so dominant that its gravity slowed the expansion of the entire universe—lost the power struggle to the rising tide of dark energy, and the expansion of the universe stopped slowing down and began to accelerate. Today on those immense scales where a cluster of galaxies is just a dot, expansion has won and space is wild. What gravity has not already bound together, it never will.

Held deep inside early halos of dark matter, clouds of hydrogen gas began falling together and igniting, becoming the first stars (fig. 64).

Inside the galaxies, protected by gravity from the forces of dark energy outside them, wonderful and complex events were and have been happening. Generations of stars produced a rainbow of exotic atoms—new chemical elements that might eons later make life possible on planets that did not yet exist. After nine billion years, just at the era of cosmic balance in the transfer of power from dark matter to dark energy, our sun and its planets formed about halfway between the center of our Galaxy and the visible edge. It might have looked like figure 65.

No sooner had Earth begun to cool than it acquired microbial life. Stars like the sun burn for ten billion years or so, and by the midpoint of its lifetime, which is now, the microbes on Earth have evolved into countless species, including one with the intelligence and technology to discover the universe, decode its history in still-arriving ancient light, and begin to fathom the meaning of our cosmic place (figs. 66 and 67). But with the same intelligence and technology, that species has overrun its planet and despoiled much of the surface, including the oceans, while scarcely understanding the implications of what it has been doing because, among other reasons, it has not yet learned to think in large-

Fig. 64. A forming star

Fig. 65. A forming planet

enough terms. It has not yet understood the necessity of balancing expansion with contraction—of accepting that this is how our universe works.

Our New Origin Story—Stage Zero

Stage One of our origin story began with the instant of cosmic inflation. But what came before that? What caused cosmic inflation? For many people, only the word *God* can answer the question of what came before. Science's approach is different: to keep pushing the beginning earlier. If there is ever a reason in principle why earlier is unknowable, we want to understand the principle.

There is a finite distance back in time that cosmology can explain with the Double Dark theory, the Big Bang, and the theory of Cosmic Inflation; but that is the end of the line for now. From the moment of cosmic inflation forward in time to today is science, but what happened before that moment is the subject of theory without evidence, and theory without evidence is metaphysics. *Metaphysics* is sometimes defined as a branch of philosophy that tries to explain the ultimate nature of reality, but we are using it quite literally to mean an area of inquiry "beyond physics"—at least for now. But although the source of cosmic inflation may be an area of metaphysics, it's not guesswork: it's based on serious calculations. Are these calculations just mathematical speculations about the real universe? We don't yet know. It's not even clear how we could test them. But an origin story must go as far as one can imagine—in fact, expand what one can imagine—if it's going to be empowering. The best theory we have, which is not to say it is the best we could have, only the best we do have, says that when we extrapo-

Fig. 66. A pensive gorilla

Fig. 67. Auguste Rodin's *The Thinker*

late the mathematical equations of cosmic inflation backward to find its source, the most likely possibility is that inflation may have been going on forever and is still going on almost everywhere but here in our universe. We warn you that without evidence this prequel is not yet officially science, but the truth is also that serious scientific research is currently being done in this field; it is mathematically sophisticated, conceptually exciting, and published in the most prestigious journals. It is a scientific extrapolation of well-established scientific ideas to see where they lead as we go backwards in time.

According to the still-controversial theory of Eternal Inflation,[2] there are two possible states of being: space-time and inflation. The realm of inflation is called the "superuniverse" (or "multiverse" or "meta-universe"). Once this state of being exists, it continues forever—but tiny pockets or bubbles form in it, which become big bangs that evolve into universes that may have laws of physics different from ours. Our universe would then be one among uncountable bubbles of space-time in the cauldron of eternal inflation. Eternal inflation is hot and dense, and the expansion of the space between bubbles speeds up exponentially forever, so that in between universes nothing can ever form and only the laws of quantum physics apply.

In one version of eternal inflation, the superuniverse is a kind of Cosmic Las Vegas—that is, the laws of chance rule. Imagine that coins are constantly flipping (fig. 68). Heads means that the coin doubles in size *and* there are suddenly two of them. Tails means that the coin shrinks to half its size. These asymmetric odds always favor expansion. Now suppose that a particular coin has a run of tails. Simply by chance it keeps shrinking again and again. Eventually it becomes so

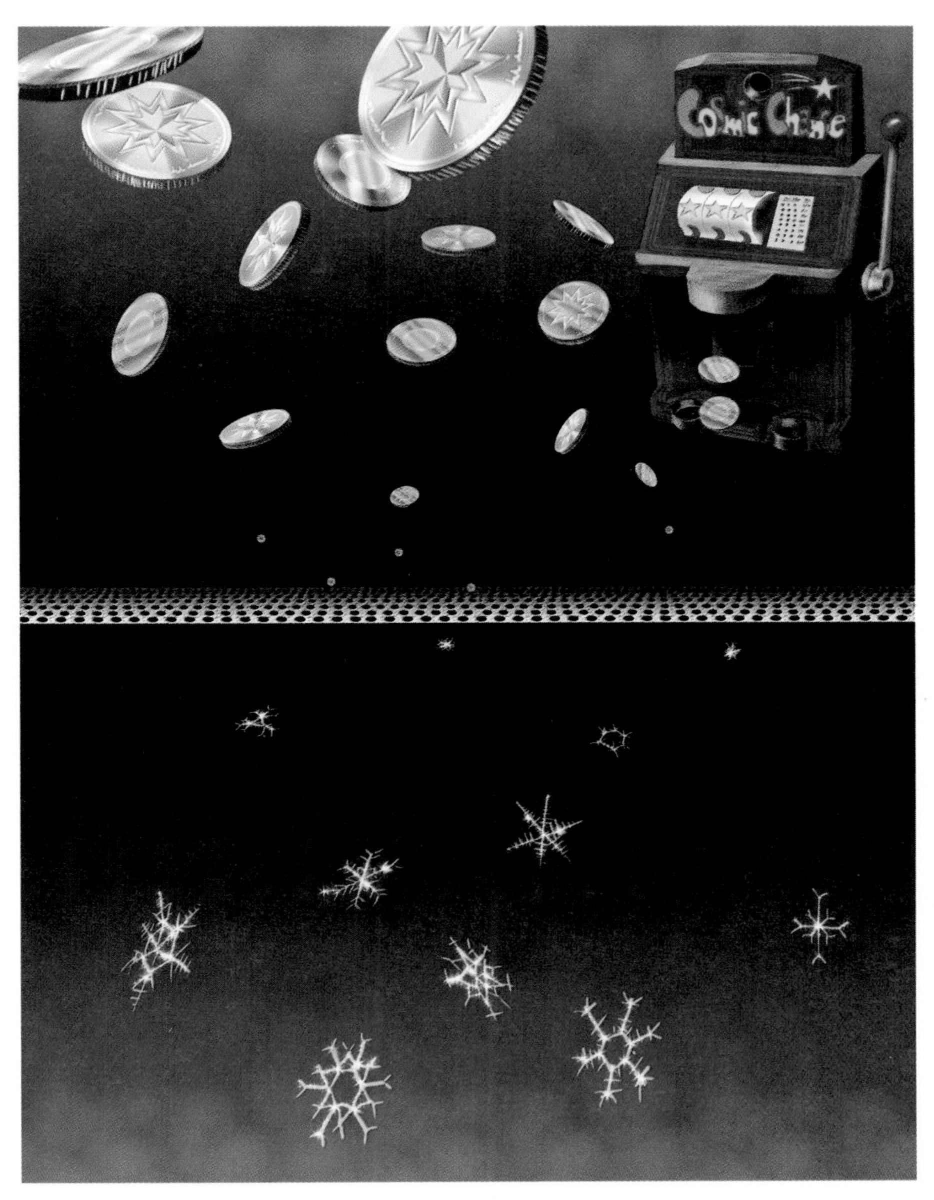

Fig. 68. Cosmic Las Vegas

small that it falls through the grating on the floor. The grating represents the instant of cosmic inflation. As the coin passes through the grating, it exits eternity. Its dark energy changes in that instant into ordinary energy. "Time" begins with a Big Bang, and a universe starts evolving. The snowflakes in the image represent universes: each will be unique, depending on the quantum fluctuations that happened to be occurring as the universe passed through the grating. Most may have laws of physics that could not permit intelligent life to exist.

If the theory of Eternal Inflation is right, then our universe—the entire region created by our Big Bang—is an incredibly rare jewel: a tiny but long-lived pocket in the heart of eternity where by chance exponential inflation stopped, time began, space opened up, and the laws of physics allowed interesting things to happen and complexity to evolve.

In ancient Egypt the gods had created the world in the midst of the Primeval Water, and the world was a bubble still surrounded by that water. In the Middle Ages the stars were affixed to the outermost sphere, and outside the sphere was heaven. Water doesn't have the inflation aspect, nor does heaven, but the idea that our entire universe is a bubble immersed in a strange and eternal state of being is a very ancient one.

▣ In figure 69 we show a second, different version of eternal inflation. This is a frame from a video that visualizes eternal inflation based on the equations that describe it, though of course it could never actually be seen. But all stories, especially highly imaginative ones, benefit from visual imagery. We have used essential visual conventions to portray eternal inflation in the video, like having the light appear to come from a source in a specific direction and having objects that recede into

Fig. 69. Many universes in eternal inflation

the distance appear to get dimmer. These conventions are used to pull us viewers into the video and are not to be taken literally.

There are many universes coming into existence within the state of eternal inflation. Each universe expands at the fixed speed of light, but the space between it and other proto-universes expands exponentially. In other words, the space starts expanding slowly but soon reaches speeds unimaginably faster than the speed of light. Consequently, the only way two universes could ever come into contact is if by chance they form so close to each other that by expanding at the speed of light they collide before the space between them has had a chance to expand faster.

The equations of the theory were allowed to run this simulation, and bubbles did occasionally form so close together that they collided before they could move apart. If our universe is one of these rare bubbles that did collide with another, astronomers may be able to see the effects of the collision by a careful study of the Cosmic Background Radiation, especially using observations by the new Planck satellite. This would be the only way anyone has suggested so far to test the theory of Eternal Inflation.

The idea that reality itself could be ultimately governed by the laws of chance disturbs many people, partly because they assume, sometimes unconsciously, that the moral order is something divine, that without it we would be animals, and that morality could not have come from what was ultimately accident. But that's exactly what evolution means: whatever simple materials are given to time as resources, even just particles and energy, time will use, and complexity, possibly in the form of life and even moral life, will emerge. "From so simple a beginning," Darwin wrote, "endless forms most beautiful and most wonderful have been, and are being, evolved."

Suppose it's true that eternal inflation preceded cosmic inflation. Why stop there? Where did the state of eternal inflation come from? Although scientists can, and do, concoct theories about this, it may be that it cannot be known, because such events, if they can be called that, may be lost *in principle* in quantum uncertainty. *Knowledge depends on the preservation of information.* The heat radiation of the Big Bang has preserved information about how our universe looked in its infancy, a mere four hundred thousand years after the Big Bang occurred, and this information lets us confidently peer back that far and then extrapolate a good deal farther. But beyond the universe created by our Big Bang, the state of eternal inflation would be a pure quantum regime in which nothing persists. Therefore knowledge would not exist. We can't even imagine how we could find out about a "beginning" for quantum uncertainty, if that is even a meaningful concept. Is this then at last the place to credit God as the literal first cause? That's an option. But rather than skipping lightly over eternity itself to paste in the idea of God "causing eternity," we might do better to think of the beginning as being just as unknown as the distant future and ourselves as true explorers, moving outward from the center in both directions. In cosmology both the distant past and the distant future are in a real sense ahead of us, the one waiting to be discovered, the other to be created.

A Transcendent Origin Story

Origin stories from cultures around the globe have fallen mainly into three categories, depending on their view of time.

> 1. The world is *cyclical* (it continually changes in the short term, but in the long term the cycle itself repeats eternally). This is the Hindu version, for example.
> 2. The world is *linear* (it's always changing, and time goes in one direction). The Bible may have been the first to adopt this view, with a creation story followed by historical heroes who each play a nonrecurring role at a particular moment.
> 3. The world is *eternally unchanging* (although if created, it went through changes in a distant, irretrievable past).

The new cosmology reconciles these ancient, deeply conflicting ideas about time by revealing that all three modes of storytelling are correct—but they apply on different scales. On the size scales of Earth, which have shaped our human minds and intuitions, the seasons are cyclical, and so are the births and deaths of generations of living beings, the movements of the planets, and the return of comets. On the size scale of the Big Bang and cosmic evolution, the universe is changing in one direction: it's expanding faster and faster, and we know of no good reason why it would ever contract. But if the theory of Eternal Inflation is right, then on that grandest scale yet conceived, the superuniverse is eternal and unchanging. Quantum possibilities burst endlessly from every sparkpoint, some of which will even become universes, yet *as a whole* nothing changes.

This multiscale view of time challenges the traditional order of storytelling. Origin stories have traditionally been told from beginning to end or from beginning to the present. But that may not be the best way to communicate the nature of a Meaningful Universe. The new origin story may start with the present and move both forward and backward, just as the Cosmic Spheres of Time move outward as the past zooms away from us in all directions and the future flows toward us at the speed of light. From the center outward is the way scientists have actually learned about the past—first the recent past: the planets and local stars, then nearby galaxies as they were only a few million years ago. Then, as instruments improved and we could see farther out into space, we saw further back in time, observing distant galaxies as they were billions of years ago. Our story is continually expanding and changing, restless as human curiosity, and this helps prevent the calcification of any version into dogma.

Some people react negatively to the very possibility of a single story shared around the world, as though there were no difference between mental dictatorship and scientific truth. But the difference is fundamental. When human beings serving their own interests dictate to us what we should believe, we must rebel; but when nature reveals some truth, rebellion against it merely sabotages our own future. Today such misplaced rebellion against nature's own revelation, as though it were nothing but an opinion, could sabotage the entire human species by dismissing the only story that may provide what Joseph Campbell said we most urgently need: "a mythology of this unified earth as of one harmonious being."

Some fear that a single origin story, even if it were true, would somehow impose a single way of thinking on everyone. But that's mistaken, too. Humans are endlessly diverse, and this is our great strength. The Cosmic Uroboros shows us that events on different scales are controlled by different laws and thus require different ways of thinking. What this means for us humans is that we can preserve kaleidoscopic diversity on the scale of our local lifestyles while still finding consensus about events on the encompassing scales of the planet and the universe.

Those who insist the universe is only a few thousand years old often have no compunction about ending it shortly. In fact, for a substantial number this symmetric sort of closure gives their life meaning. But it's not a meaning that has anything to do with reality, that can ever serve a global civilization, or that can ever support peace and a shared commitment to our descendants and our planet. Since human consciousness looks outward from the center, as we discussed in chapter 5, a tiny consciousness of history is mirrored in a tiny consciousness of the future. The end result is a consciousness too small to notice, let alone appreciate, most of reality. The larger the past that our minds encompass, the larger the future we become capable of imagining, taking seriously, and protecting. In this way the history of our universe may be the key to our future.

The origin story of the new universe also challenges deeply held assumptions about what kind of story can actually satisfy spiritual longings. If we define *spirituality* as "experiencing our true connection to all that exists," then the new origin story comes closer than any other to helping us fulfill that longing. There is no sense in either judging the scientific story (unsatisfying! too hard to get! too foreign to care about!) or dismissing it (just a theory, has nothing to do with God's creation). What makes sense is to seek passionately to absorb these discoveries, using every cultural tool available to bring home to ourselves this new knowledge of where and what we humans actually are. If anything is a miracle, this is it: that precisely at this pivotal moment for our species, when so much is required of us, a cosmic opportunity has fallen into our laps. A potentially empowering, transcendent origin story has appeared that could unify so many around the world who may not see eye to eye on many other things. All they need to agree on is that our place in the universe is extraordinary and that humanity could have a cosmically long future. This level of agreement could change the world.

Chapter 8
Cosmic Society Now

Building a cosmic society is not a dream for the distant future, like galactic travel. It's about today. In fact, the future world is far *less* likely to achieve a cosmic society if we don't begin moving in that direction now. Because of the speeding-up nature of current exponential growth, the longer we procrastinate in dealing with the tangle of global problems, the wider the tragedy will be. Right now humanity's overriding need is for a transculturally shared vision for how to solve global problems, and a cosmic society is the only serious candidate we know for an organizing principle that would allow such a shared vision to flourish. Any program of action not based on the universe as we now understand it is doomed.

History shows that having a shared cosmology, even a wrong one, can unify and inspire a culture, but the very power of a cosmology means that when it gets overturned this event can shake up the most fundamental institutions of society. This is in fact what happened four centuries ago in the last cosmological revolution on the scale of today's, when the science of Copernicus, Galileo, Kepler, and Newton overthrew the medieval cosmology of the heavenly spheres. The collapse of that cosmology undercut the rigid social and religious hierarchies of medieval society, which had been justified by the supposedly hierarchical order of the heavens. Soon the divine right of kings itself was challenged, and kings of first England and then France lost not only their thrones but also their heads.

It's indisputable that practical consequences can result from a change in cosmology, but no one can predict what they will be. This is why it matters how scientists, particularly, present the story. No scientist should ever claim that science is the last word in a decision that could shape the larger society. Science is the first word. It's the foundation. Everything else needs to build on it. Thus the crucial relevant concepts of cosmology and the other historical sciences like geology and evolutionary biology must be made understandable to everyone.

Our species stands at the turning point in a struggle between despoliation of our planet and a big new perspective. As Martin Rees, the leading astrophysicist and cosmologist who is past president of the Royal Society (Britain's national academy of sciences) has written, "We've now entered a unique century, the first in the 45 million centuries of Earth's history, in which one species—ours—could determine, for good or ill, the entire planet's future."[1] But unlike the power struggle between dark matter and dark energy, the struggle between despoliation and the new perspective will be determined not by the laws of physics alone but by what human beings believe about ourselves and each other and our collective potential. This is why we need a shared story. The scientific origin story delegitimizes the forces of despoliation by giving us a big new reality in which our dynamic planet, rich with life, is cosmically rare and invaluable. It tells us that there is an immense future ahead of us and our descendants. But we must first make a successful transition during the next few decades from exponentially increasing interference with Earth's natural systems to a sustainable relation with our remarkable home planet.

Perhaps it will help if more people appreciate how extraordinary Earth is from a cosmic perspective. There is a commonly held notion that Earth is an average planet of an average star, but nothing could be further from the truth. The first planets outside our solar system (extrasolar) were discovered in the mid-1990s, and by this writing nearly five hundred extrasolar planets have been discovered. But none of them appears to be suitable for life. The more we learn about our solar system compared to other planetary systems, the more special Earth appears to be. There are at least six ways that Earth is an unusually suitable planet for complex life.[2]

First, many of the extrasolar planets that astronomers have discovered thus far are "hot Jupiters," that is, massive gas giant planets in orbits very close to their stars. These planets most likely formed farther out and spiraled inward, destroying any small earthlike planets on the way. But our Jupiter, which is by far the most massive planet in the solar system, is still far from the sun.

Second, many massive planets that are not hot Jupiters have very elliptical orbits in other planetary systems, coming in close to their star and swinging out again, but our Jupiter traces a nearly circular orbit far enough from Earth not to disturb it but close enough to stabilize Earth's orbit. Jupiter also helps to protect Earth from frequent collisions with comets or asteroids that could have prevented the evolution of complex life.

Third, Earth is the only planet in the solar system that has been in the sun's habitable zone for its entire existence, close enough so that there is liquid water on its surface, but far enough that water won't evaporate and ultimately be lost. Moreover, the sun is a long-lived star whose luminosity is very stable.

Fourth, Earth's thin crust and abundant surface water allow continued geological activity, which recycles carbon and other elements essential for maintaining life.

Fifth, the moon, which was created by a Mars-size protoplanet crashing by chance into the proto-Earth, stabilizes Earth's rotation and climate.

Sixth, our entire solar system lies in the "galactic habitable zone"—not so close to the dense galactic center that dangerous radiation threatens life, but not so far from the center that there's insufficient stardust to make rocky planets.

In recent years, a seventh way that Earth is highly unusual may have been discovered. Most star systems have much more dust and debris among their planets, which may mean that any earthlike planets around these stars would suffer much more comet bombardment than Earth, causing frequent extinction events that could impede the evolution of advanced life. Why is our solar system so clean? Astronomers working at the observatory in the French Mediterranean city of Nice have recently discovered a solution. This "Nice model" starts with the three outermost planets—Saturn, Uranus, and Neptune—initially much closer together than they are now, and much closer to Jupiter, with Neptune closer to the sun than Uranus (the reverse of their order now) and a giant debris belt of cometlike objects beyond Uranus. It is quite plausible that they formed this way. Computer simulations of the Nice model show that gravitational interactions among the planets would then prevent Jupiter from spiraling inward, and after about eight hundred million years push Neptune beyond the orbit of Uranus and out into the debris belt. When that happened, some of these debris objects would have been shunted inward toward the inner planets, while most would have been ejected to great distances from the sun, becoming the Oort cloud of very distant comets. But enough of the debris would enter the inner solar system to explain what astronomers call the Late Heavy Bombardment, which lasted from about six hundred to eight hundred million years after the formation of the solar system, causing a large number of the impact craters on the moon, and probably also many impacts on Mercury, Venus, Earth, and Mars. Life could develop on Earth only after the bombardment ended.

Infrared telescopes in space have recently given us additional important information about this by detecting debris disks around nearby stars. (NASA launched the Spitzer Space Telescope in 2003, and the European Space Agency launched the Herschel Space Telescope in 2009.) Astronomers were surprised to find that our solar system has much less dust and many fewer comets than the planetary systems of most stars like the sun. The Nice model intriguingly explains this by the gravitational ejection of the vast majority of the material in the original debris belt to great distances or entirely out of the solar system, as a result of the migration of Uranus and especially Neptune. Information about planets around other stars is rapidly becoming available, so we can expect to learn much more soon and possibly discover even more ways that Earth is special. The better we grasp this, the clearer it becomes that despoliation of this planetary jewel is a cosmic crime.

There is a second common notion holding us back that is also incorrect and self-destructive: the notion that we cannot protect the Earth unless we stop all growth, causing our economies to collapse. It was once thought to be an economic law that energy use rises with economic output. But starting with the "oil shock" of 1974, many countries broke this link. Global and US carbon emissions per capita peaked in 1974 and have remained fairly steady since then, rising in developing countries while they have declined by a third or even more in several European countries such as France and Germany and in a few US states including California.[3] Yet the standard of living rose in these places where the carbon emissions per capita declined.

It will be a challenge to break the link between economic output and environmental damage in developing countries, so that billions of poor people in the world can benefit from continued economic progress. No long-lived society can be based on warfare between environmental protection and social justice. The solution involves learning how to produce and use energy more intelligently and a widespread commitment to change.

People can steadily improve their lot in many ways without using more resources or causing environmental harm; this does not conflict with any law of economics or physics. It will involve making subtle changes in both lifestyle and technology. For example, since people living in cities use fewer resources per capita, it is good that the world is becoming increasingly urbanized. Steady improvement in electronic communication can increasingly substitute for physical travel, and more and more people are already choosing to work from home or conference call rather than travel for meetings. The capital costs of solar energy are rapidly decreasing. New smart grids can move electricity safely and cheaply. But these things won't happen fast enough unless the economic environment is changed so that we stop subsidizing fossil fuels and other harmful technologies and start investing enthusiastically in sustainable alternatives. The first industrial revolution was the replacement of human labor with machines, often powered by fossil fuels. But continued growth in the use of fossil fuels and other resources will result in a world that is increasingly hostile both politically and physically. The next revolution must be toward environmentally friendly growth.

For this to happen in democracies, it will probably be essential for an increasing fraction of the public to accept the best scientific thinking. H. G. Wells wrote in 1919, "Human history becomes more and more a race between education and catastrophe." It's not clear that education is winning this race. Sadly, the United States population, which has produced so many great scientists, is near the bottom in a list of countries in accepting biological evolution.[4] Even though evolution is the central unifying theory in biology, supported by overwhelming fossil and genetic evidence, and the teaching of creationism has been judicially prohibited in US public schools for the past quarter century, fewer than half of Americans responding to opinion polls agree with the statement that humans and other living things evolved over time. Many who do not accept evolution assume that it is incompatible with their religion, even though representatives of many religious denominations have stated that biological evolution is consistent with their beliefs.

Similarly, despite authoritative reports by the most reputable scientific sources, many Americans are still not persuaded that human-caused global warming is a reality. This is probably both because they fear that having to do something about it now could threaten their economic well-being and because of the pervasive but unspoken sense that we're insignificant motes, so we can't possibly have made such an impact on the entire planet. But it is unquestionably due in large part to the fact that the news media have allowed self-interested parties to muddy the waters. Remarkably, a small number of scientists have used prestige that they earned in other areas to obscure the truth again and again, using virtually the same strategies on numerous issues from acid rain, tobacco smoke, and the ozone hole to global warming.[5]

What is needed is not education on details but rather a change in people's basic cosmology—their whole sense of how the universe came to be and how it works. On both the national and the world scales, this is probably a prerequisite for cooperation on changes that can make us all healthier, more secure, and in the long run more prosperous than we otherwise would have been.

Different people will grasp the scientific part of the new universe story at different levels of technical detail, but its essence is simple enough for a child to understand: we are stardust evolving into awesome complexity over billions of years in an expanding universe shaped by invisible dark matter and dark energy. Our planet is very special and maybe unique in all the cosmos. Our ancestors are not just our grandparents and great-grandparents, not just our ethnic group, not just the human species, not just life or even Planet Earth, but stars, galaxies, dark matter, and all the forces of nature, living and not, in an unbroken chain back to the Big Bang.

If, by knowing who we really are in this way, we early-twenty-first-century earthlings begin to develop a cosmic society with the goal of a long-lasting global civilization, it is likely that some of our distant descendants will move outward into the Galaxy and could eventually radiate life and intelligence throughout it. Since our Galaxy (eventually Milky Andromeda) will be the future visible universe, our descendants could become the source of intelligence for the entire future visible universe. Even if there are already aliens contributing to the intelligence of the Galaxy, we and they are each the beginning of a unique kind of intelligence, shaped by a unique evolutionary path on a unique planet. The cosmos is richer for having all of us, even though none of us may yet know of one another. We are all together in the glowing Eye at the top of the Cosmic Density Pyramid. Stars are still forming in the Milky Way, and some will be burning for trillions of years. There is no reason that stardust-based life could not go on that long and evolve to levels we can no more imagine today than a toad could read this book.

The future of intelligence in the visible universe could thus depend on humanity—and humanity's survival may depend on those of us who are alive today. We have to admit the disastrous course of business as usual; invest in scientific research, including in the social sciences, to find all possible openings; agree to a great extent on what can be done; negotiate in good faith as to who will do what; and stick with the overall plan through thick and thin, despite inevitable short-term crises. To uphold this worldwide goal must become an article of faith and of honor just as serious and indeed sacred as upholding the cosmos was to the ancient Egyptians. Earth is negotiating with us right now and waiting, not so patiently, for a good-faith reply. If humans don't make it, the universe has plenty of time and space to try and evolve intelligence again, but humanity will be sloughed off as if we had never been.

A cosmic society could be the Ark that carries all of humanity, of every color and kind, from a bleak future on a declining planet to a cosmic perspective and to the safer shore of a stable, long-term civilization. It would not take a majority at first—not even near. As Margaret Mead is supposed to have said, "Never doubt that a small group of thoughtful, committed people can change the world. Indeed, it's the only thing that ever has." A small group can start building this Ark, but everyone has to be welcome on it. This may be what it takes to ensure the future not only of humanity but of intelligence in the visible universe. How fortunate that the cosmology of a Meaningful Universe, which may make possible a civilization that in the long term will explore the Galaxy, is the very same cosmology we need right now on the most down-to-earth level.

Some people oppose the idea of space travel and wouldn't even want humanity to be a template for intelligence on other worlds; they fear that if our human, or more likely humanoid, descendants have the chance, they will run amok and wreck other planets the way European conquerors wrecked so many indigenous cultures that they colonized. But that's the wrong fear. There's no need to worry about that. Space pioneering would be impossible for the short-sighted, egocentric kind of people we were and in many cases still are today. To explore and gradually move out into the Galaxy is a project that could be successfully undertaken only by a long-lived civilization with a shared, unifying cosmology that accurately reflects the universe. The civilization would have to be stable enough to welcome home space travelers or their descendants even generations later. To be able to achieve this, the civilization would have to be a cosmic society, understanding our central place in the universe and the value of intelligence to the evolution of the whole. This outlook is completely inconsistent with narrowness, greed, and willful ignorance of the Other—traits central to the plunderer mentality. So the good news is that probably the only kind of creatures that could successfully pioneer the Galaxy are ones that should.

Maybe this is why no aliens are here. They may have the technology but not the mythology. Maybe intelligence of the kind that develops mathematics and high technology is relatively common throughout the universe but what is vanishingly rare is creative intelligence of a deeply artistic and philosophical bent: that seeks harmony, rather than self-aggrandizement; that conceptualizes its place in the cosmos as precisely and accurately as it can; that develops mythic imagery that sweeps its members up imaginatively and lets them experience their cosmos through daring metaphor. This kind of human intelligence is what created the mythologies that bind religions and nations together—but no such mythology yet exists that can bind our species together despite religious and national differences.

A common universe can provide common ground. We humans are so diverse that the way to deal with global problems is not to impose global solutions but to *cultivate the common ground of a large-scale vision based on principles* and to encourage small-scale, decentralized solutions, appropriate to different situations, created by all kinds of people who are inspired by that vision and its goals. René Dubos, the French-American microbiologist and ecologist who coined the phrase "Think Globally, Act Locally," pointed out that as more public activities and experiences become globalized, the countertrend of identifying with a chosen neighborhood or community—what he called "local patriotism"—will increase, since people seek the comfort of small-scale communities. This is exactly as it should be. We can learn to think on multiple size scales, deciding which is appropriate to the matter at hand.

Enlarging Human Identity

Each of us is an entire uroboros with roles to play on multiple size scales (fig. 70). We each have an individual self-consciousness, represented by the tip of the tail. On increasingly larger scales, we identify ourselves as part of a family, of a tribe and/or community, and of a religion and/or nation. But our deepest identities are far larger and much older than any culturally invented religion or nation: we are human beings, living things, earthlings, and at least part of the self-consciousness of the universe. The moment enough of us recognize this—and become willing to accept the logical consequences—is the moment we become a cosmic society. It's our generation that needs to take the leap, collectively, from religion or nationality as the deepest level of identity to the species level and see ourselves above all as human. All human beings are closely related (more closely, in fact, than the members of any other species to fellow members of that species). It's the human species that merits our deepest loyalty, and anything that threatens the survival or health of our species threatens each of us.

This is not an easy transformation. It seems at first to conflict with our narrow tribal instincts, our learned priorities, even our love of irony and cynicism. It entails a major change of thought. Yet it is easy because it doesn't take capital or labor. We are all conflicted: we want something easy, but we also crave something big and important, something life-changing—and that can't be easy. But once people have actually tried on our species identity and found how illuminating it is

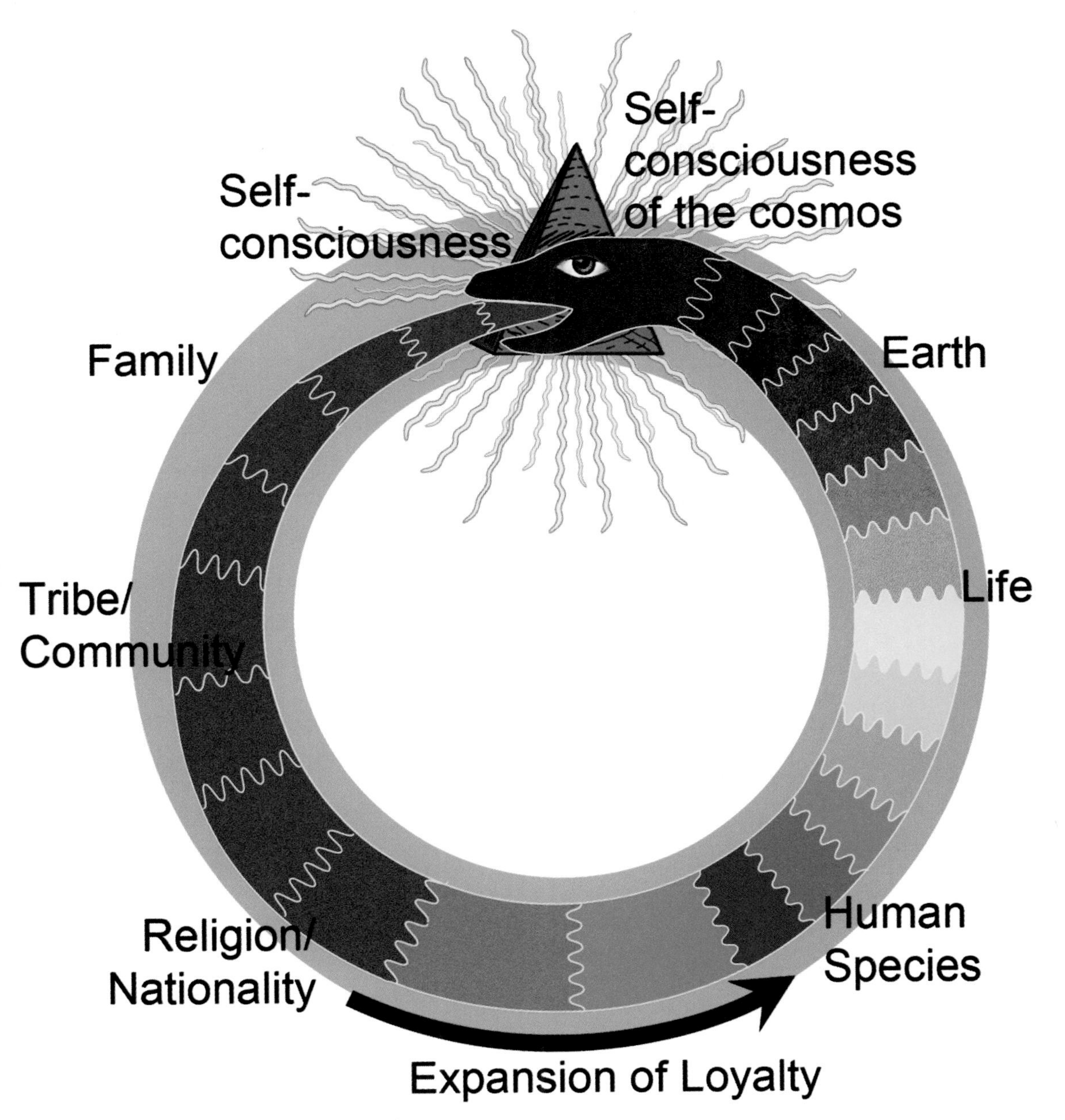

Fig. 70. The human identity uroboros

and how empowering it feels, expanding their sense of identity the rest of the way to life, Earth, and cosmos will be easier. These newly discovered identities coexist within each other, waiting for us to discover them and recognize ourselves.

Two bricklayers worked side by side in medieval France. When one was asked what his work was, he replied glumly, "I lay bricks, one on top of the other, again and again." When the other was asked, he said proudly, "I am building a cathedral to the mother of God." Neither of the bricklayers was wrong, but that does not make their answers equivalent, because the consequences for their lives were so different. So many people today think we're all just laying bricks and that it's grandiose nonsense to project meaning into the bricks. We are all so intimidated by the reward structures of our jobs and of our short-attention-span society. But at this pivotal point in the entire evolution of our species, we need to crack open our imaginations! We have unlimited imagination, but we're not using most of it. We too are building a cathedral.

Something of mythic importance is being demanded of us by the accident of being born at the turning point.[6] To understand what this means for our twenty-first-century lives, the techniques that all earlier cultures employed to share and develop their cosmologies will be essential: language, imagery, story, ritual, mystical experience, and other artistic expressions will all be important to help the new scientific cosmology become comprehensible beyond a narrow circle of experts and meaningful for us all. And so such techniques must flourish in a cosmic society. But like the bricklayers' explanations, not all accurate attempts at explaining the new universe are equal.

"Romantics are made of stardust," goes an astronomy joke, "but cynics are made of the nuclear waste of dead stars." Yes, both images are scientifically acceptable, but why should anyone fight to save humanity if they see us as just nuclear waste? If you were a child, would you rather learn that you are stardust or nuclear waste?

The choice of metaphors to portray the new universe must be strategic. Any particular metaphor will have a certain kind of impact on people—on their hearts and minds—and that impact is just as much a scientific fact as the discovery that the metaphor was trying to communicate. Every cosmological metaphor is a double-edged sword: it is so important to make sure that, as with the nuclear waste metaphor, we're not cutting ourselves with one edge while admiring the other.

For example, several professional cosmologists have taken the position that the new astronomical discoveries are bad news. Their reasoning starts from the possibility we explained earlier: that in tens of billions of years dark energy may inflate the space between our Galaxy and all the distant galaxies so much faster than the speed of light that those galaxies will disappear from the visible part of the universe. They conclude that since Milky Andromeda will be left alone, this is "the worst of all possible universes." It's a catchy phrase and one that plays well in a culture of cynicism, but it's a self-sabotaging interpretation of our new scientific knowledge, and even cynics (underneath their layers of armor) want to live and flourish.

The truth is that the distant future, like most of the distant past, is unknown. We become our own enemies if we let our emotions and attitudes toward life today be dampened by such outrageously distant speculation. Permanent changes, including galaxies disappearing over the cosmic horizon, are in the nature of an evolving universe—not something to be mourned, and certainly not billions of years in advance. Rather than feeling sorry for our very distant descendants and harshly judging our entire universe today based on what it may be like for them, we might as well appreciate our cosmically central place and the huge power the universe has effectively entrusted to us to influence the distant future. We benefit much more from focusing on the center, our time, and extending it outward as far as knowledge permits. Rather than concluding that this is the worst of all possible universes, let's realize that if we protect and nurture Earth and we don't destroy our descendants' chances to expand beyond it, then we humans—with all our flaws and all our dreams—could be the beginning of the whole future of intelligence in what will ultimately be the visible universe. And our descendants will have at least hundreds of millions of years to enjoy a hospitable Earth—at least a thousand times longer than our species has existed so far.

There is no dismissing "worst of all possible universes" or "nuclear waste of dead stars" as "mere" metaphor and thus not to be taken so seriously, because with respect to the universe as a whole, there is nothing *but* metaphor.[7] Language was invented to describe earthly things and earthly experiences and is always metaphorical when applied to the universe as a whole. If taken literally, scientific cosmology is completely misleading. There was no loud bang at the Big Bang, and it wasn't big. (There was no size to compare it to.) *Metaphor is our only entrée into invisible reality.* Not understanding this, some people, in a misguided attempt to defend science from imprecision, argue that metaphorical descriptions are unscientific. But metaphor is inevitable in human understanding, according to the sciences of how the brain works. It is fundamental to the way we think, create, and even discover. So the metaphorical nature of science is not a dirty little secret but an opportunity for creativity, not only in developing scientific theories but in the thought processes of the entire culture. We must figure out how to grasp these cosmological concepts or they will never serve us. But there's no need to reinvent the wheel: the creative use of imagery, metaphor, and story to make people feel they belong to their universe has been a central tool of human cultures for thousands of years and will continue to be. It's only our understanding of the universe that has changed.

Education in a Cosmic Society

Childhood is the age when most people acquire the picture of the universe they will carry in their intuition all their lives. How wonderful it would be to teach children the real story as early as possible, at whatever their level of comprehension (fig. 71). Today's children could be the first generation ever raised in the universe they actually live in. Too many people tell silly stories to children: a cartoon that recently appeared in the Sunday comics showed a child being told that at the end of the day God puts the sun to bed and covers it with a blanket of night. This hearkens back to ancient Egypt, where it was thought that the goddess Nut (the heavens) swallowed the sun every evening, and the sun was dark as it traveled through her body and became bright again when she gave birth to it in the morning. Another example of counterproductive silliness is the supposedly inspirational motto on greeting cards often sent to graduates, "Shoot for the moon—even if you miss, you'll land among the stars." The notion that the stars are not far be-

Fig. 71. Child and cosmos

yond the moon hearkens back to the Middle Ages, when the stars were thought to be fixed to a sphere that rotated around the earth every twenty-four hours. We are among the stars right here! Communicating a silly picture of the cosmos may seem as harmless as pretending there's a Santa Claus, but it's not. Every child eventually figures out the truth about Santa Claus, because there are practical consequences. But with the cosmos, many people will never face that moment of truth.

It's also detrimental to teach children that "the universe is so vast and we're so tiny." That may sound more scientific, but it's wrong. We're at the center of all possible sizes: there are things as small compared to us as we are compared to the visible universe. To understand our central place on the Cosmic Uroboros, all that children really need to understand is the powers of ten—a concept that is surprisingly easy to teach to elementary school children. The powers of ten are exciting to them. We have taught it to second and third graders and seen how fast they can catch on to this shortcut that bursts open their consciousness to jaw-dropping sizes they had never had the tools to imagine.

Climate chaos, extinction of species, and other looming problems can be terrifying to children, since they sense—and rightly so—that the messes are being left to them. It's our job to give them both truth and hope based on a realistic sense of empowerment; the alternatives to empowerment are too often selfishness, cynicism, or despair. It's also our job to teach them how far to believe and when to maintain skepticism, and this requires that they be taught the crucial role of evidence. Education in a cosmic society would teach children *to function intuitively in the real universe*—to believe in the real universe, not just memorize facts about it and tack those facts to the outside of their brains like sticky notes while their intuition goes unchallenged and unchanged. A cosmic education would be a source of confidence and wisdom and a unifying outlook among the young as they confront the chaos that they are inheriting. It's our job to teach them, also, that Earth itself is not a mess but a jewel of the cosmos, rich with life and potential, and possibly unique in all the heavens. We may be the locus of the universe's consciousness. If we could just stay aware of this—if our thinking lived up to this standard—we humans could solve our worst problems.

The way to begin teaching the truth to our children is to begin acknowledging to ourselves how far-reaching the consequences of today's decisions may be. Cosmological time is what we are part of, whether people believe it or not. Political decisions no one is paying much attention to today can silently prune out entire worlds that might have been, on every size scale from depriving a child of an education to wiping out intelligence in the Galaxy. "Cosmological time" may at first seem hard to imagine or ridiculously abstract, but it's as real as anything can be, and we need fearless imagination to figure out how to present it to each other—accurate and alive, with effective metaphors that expand the consciousness and captivate the young, so that our culture will begin to think in cosmological time and experience its influence.

Many people will remain in denial as long as the society around them lets them. The default position is to assume that a change as big as the one we are discussing is unrealistic, but the fact is, no one can say what is or isn't realistic unless they know what universe they're living in. This is an age for heroism—for people willing to start *believing the evidence* that we are at the center of a new universe and at a pivotal moment for humanity, and that we must act accordingly. The realists are not the ones arguing against change: the realists are the ones who are helping make the real universe believable.

It's now possible to feel part of the new story and to want to live fully and responsibly in cosmological time, but it's not possible for anyone to do this alone. Reality is always a social consensus. It's a communal perception. A person confident of a reality no one else sees is considered insane. And thus to achieve a widespread cosmic society there needs to be a founding group—small at first—of willing humans from anywhere and any background on the planet to start this crucial, creative work. It might call itself the Cosmic Society. It would strive to open people's minds, with the goal of anchoring the coming social consensus in the truest universe of our time.

It's a strange universe at first, but this is our real home. People will inevitably get used to it—but hopefully soon enough to reap its benefits while there is still time to right the mistakes we made in our earlier ignorance. A cosmic society founded in a Meaningful Universe could become for our world a wellspring of bonding energy and artistic creativity, the way ancient cosmologies served our ancestors ▣.

Could the idea of a cosmic society end up being a religion? No. It's too free-thinking. It's more like an ethic. It requires faith not *despite* the evidence, as some religions encourage, but *in* the evidence—and in the possibilities that the evidence suggests for our species. It makes possible for us the shared experience of actually belonging to the cosmos that the evidence tells us exists. Above all, it allows us a way to elevate our thinking to what our times demand. It doesn't tell us "the meaning of life." It tells us that the meaning of life depends enormously on the *scale* in which you consider it. Everyone has an opinion on the meaning of life to an individual on Earth, but what is the meaning of life to Earth? Earth, after all, has had to deal with all of it! We humans have to think about this *for* Earth because as far as we know, we're the only ones who can. We are the thinking part of Earth. To do so we have to—for the moment—think on Earth's scale. And until we discover aliens capable of sharing the mind of the universe, thinking on the cosmic scale is our job, too.

As the intelligence, or part of the intelligence, of the universe, we have a responsibility larger than our responsibility to Earth. If anything is sacred, it must be our responsibility to the universe itself. That doesn't mean we have to travel around the Galaxy fighting evil forces, as so many science fiction TV shows tell us. Our responsibility to the universe is right here: it is to protect humanity, because humanity is the guardian of an extraordinary occurrence in cosmic evolution—a brain that can conceive of the universe. Our existence matters to the universe; Earth must put up with us, but if we give our species time, we may truly change the universe.

When we push our imagination in these ways and consciously step outside our individual viewpoints into our larger roles as the consciousness of Earth and of the cosmos, we are bringing our mental world into harmony with the universe, learning to encompass multiple size scales, retraining our intuitions, and helping to create a cosmic society.

Here is a hypothetical fragment of an origin story from the distant future.

"Everyone in the Galaxy came from one little planet called Earth. The cooperative genius of a small group of ancient earthlings made them rebel against the robotic march to death that the majority was on—to use up the planet with no plan for the future. It was the acquisition in the early twenty-first century of the ability to see into the cosmically distant past and thus project far into the future that allowed the rebels to grasp the meaning of their own evolution from the first available scientifically cosmic perspective. The rebels saw the implications of what a few scientists were already suggesting but had not yet fully appreciated: that intelligent beings on Planet Earth were cosmically central, living at a pivotal moment in time, and had a potential destiny as vast as the cosmic past if they would only protect it. This awakening led to the great conversion from short-term fragmented identities to the first serious long-term species identity. Those early rebels changed the course of history and led to the Galactic community of our time."

Let's keep that story possible.

Frequently Asked Questions

1. How reliable is cosmology? It's not as reliable as physics, is it?

Modern cosmology is a historical science, like geology and evolutionary biology. The historical sciences attempt to understand not only the way the universe, the earth, and living systems work but also the historical path that led to the present. Some postmodern thinkers and many people who prefer traditional accounts of our origins claim that because the actual past was unique, the historical sciences provide a lower grade of knowledge than such laboratory sciences as physics and chemistry, which provide timeless principles and in which the effects of changing conditions can be explored in experiments. But this is a serious misunderstanding. In both experimental and historical sciences, successful theory has to both explain existing knowledge and predict new knowledge that is later actually discovered. The only real difference is that predictions of the historical sciences concern not what will happen but what will be discovered about what has already happened.

Knowledge from the historical sciences can be as reliable as knowledge from the laboratory sciences.

Modern cosmology is entirely a product of the past century. It is based on Albert Einstein's theory of General Relativity—our modern theory of space, time, and gravity—which continues to be confirmed by every experimental test.[1] Alexander Friedmann and George Lemaitre used General Relativity to predict the expanding universe, which was confirmed by Edwin Hubble's discovery that distant galaxies expand away from us with speeds proportional to their distance. George Gamow and his colleagues predicted the temperature of the heat of the Big Bang—the cosmic background radiation, subsequently discovered by Arno Penzias and Robert Wilson. The Cold Dark Matter theory predicted the size of the fluctuations in the temperature of the cosmic background radiation in different directions, which was confirmed by NASA's Cosmic Background Explorer (COBE) satellite in 1992. Several major observational programs in the late 1990s accurately measured the expansion rate and other fundamental characteristics of the universe for the first time. Since then, many observations have repeatedly crosschecked all the basic foundations of cosmology. We now at last have a reliable picture of the evolution and structure of the universe. However, there are still many basic things that we don't yet know, such as the nature of the dark matter and dark energy, or why the universe has the observed quantities of dark matter and dark energy. Science is frequently incomplete, something that my advanced students appreciate because it means that there is much left for them to do. But that does not mean that it is not reliable as far as it goes. Cosmology remains an exciting science since there are many basic facts still to be discovered. JRP

2. The Big Bang cosmology sounds like a very Western worldview. Why is the Western scientific view of the universe superior?

Science is no longer "Western"—it is now knowledge produced by scientists all over the world that is shared with all those who learn science. The reason science must be taken seriously is that it makes reliable predictions on which both technology and worldviews can be based. You trust in the fruits of science whenever you fly in an airplane. You should also trust in them when you think about the future of Earth.

The basic theory of modern cosmology, the Double Dark theory, has made extremely detailed predictions about the cosmic background radiation and the evolving distribution of galaxies that were subsequently confirmed by observations—see figures 32 and 33, for example. JRP

3. How does $E = mc^2$ explain the relationship between matter and energy?

Matter, as far as we can tell, consists of elementary particles of various types. For example, atoms consist of electrons around a massive nucleus; the nucleus consists of protons and neutrons, which in turn are made of quarks held together by gluons. We think that the dark matter is also some sort of elementary particles—and we have various theories as to what those kinds of particles might be. Energy is something different. It can be carried by particles, but it's not the particles themselves but rather something that is reflected, for example, in the speed of the particles or in other properties that the particles have. For example, the kinetic energy K of a particle of mass m and speed v is $K = \frac{1}{2} mv^2$ (as long as the speed v is much less than the speed of light).

The key to understanding the connection between matter and energy is Einstein's famous formula $E = mc^2$. The speed of light (c) is a very large speed compared to speeds of things that we are familiar with, like jet planes and rockets. It follows that a very tiny amount of matter can be turned into a huge amount of energy. And in fact that's the basis of all of astrophysics—and of life. The sun uses nuclear fusion to convert mass into energy that is then radiated, and when it's received on earth, plants can absorb it and convert it into stored energy, which we then use as food or as fossil fuels. Stars can store up energy in other forms: uranium, for example, is made when massive stars explode as supernovas.

The amount of stuff in the universe is measured in terms of density, the total amount of matter and energy in a representative volume of space (a volume large enough to hold many galaxies). Although matter and dark energy both contribute to density, they have quite different effects on the expansion of the universe. The contributions of dark energy, dark matter, and atoms to the cosmic density are shown in figures 30 and 31. JRP

4. What might the dark matter be? What are we hoping to find out about dark matter in the next few years?

We scientists are hoping that if we can discover what the dark matter actually is, that could be a tremendous clue to how the whole universe is put together. We have this modern theory of particle physics called the Standard Model, which predicts correctly the results of all the experiments yet done at high-energy physics laboratories. But the Standard Model is nevertheless tantalizingly incomplete because it doesn't give us answers to a lot of basic questions, and there's no room in it for dark matter. So physicists have for decades tried to go beyond the Standard Model. The basis for most attempts to do this has been the hypothesis of supersymmetry.

Supersymmetry predicts that all the fundamental particles that we now know have undiscovered partner particles, called superpartners. The dark matter would then be the lightest superpartner particle (LSP). In most versions of supersymmetry theory, the LSP is predicted to be stable (that is, it cannot decay into any other particles), and it is a natural candidate to be the dark matter particle.[2] Furthermore, when we calculate theoretically how many of the LSPs would have survived the conditions of the early universe, the answer is in the same ballpark as the amount of dark matter astronomers have actually measured to exist in the universe today.

The hypothetical supersymmetric partner of the electron is called the selectron, and it would have the same electric charge as the electron. The quarks would have partner particles called squarks, and the photon and the strong and weak force particles, the gluons and the W and Z particles, would have partners called the photino, the gluinos, the Winos, and the Zino. The reason that none of these hypothetical partner particles has been discovered yet is presumably because they're massive and we haven't had enough energy at accelerators to make them. There are good reasons to think that the energy that will be available at the Large Hadron Collider (LHC) in Geneva, Switzerland, will allow us to start making supersymmetric partner particles. If so, they will immediately decay into the LSP, which will leave the detectors without being detected. The evidence that this is happening will be that the LHC detectors will notice missing energy and momentum that is being carried away by the LSP. Astrophysicists Gary Steigman and Mike Turner suggested calling such dark matter particles Weakly Interacting Massive Particles, or WIMPs—an especially appropriate name for them because, although such particles may be more massive than even the heaviest atoms, they have only weak and gravitational interactions, so they mostly just go right through you, like neutrinos. Nevertheless, extremely sensitive experiments are now under way that should be able to see the rare events when WIMPs bounce off nuclei. Such experiments are done in deep underground laboratories to shield them from cosmic rays. Other experiments using space satellites such as the Fermi Gamma Ray Space Telescope are looking for evidence that two WIMPs can interact with each other and turn into other particles including high-energy photons called gamma rays. We hope that a combination of results from such experiments will soon tell us the identity of the dark matter particle—or else rule out such theories. JRP

5. How much do neutrinos contribute to the cosmic density?

The cosmic density due to neutrinos is between 0.1 and 1 percent. The upper limit comes from comparing the distribution of luminous red galaxies in the Sloan Digital Sky Survey (SDSS) with predictions of the Double Dark theory including neutrinos of various masses, and also using the latest cosmic background radiation data plus the expansion rate of the universe.[3] A more restrictive upper limit of about 0.5 percent is obtained with the latest SDSS data and slightly stronger assumptions.[4] The lower limit comes from the minimum mass of the most massive neutrino, which is deduced from measurements of a phenomenon called neutrino oscillation that occurs when high-energy cosmic rays hit the top of the atmosphere.[5] JRP

6. What is the dark energy?

The dark energy is what is causing the universe to expand faster and faster. It might just be a constant property of space itself—the more space, the more dark energy—which causes space to expand exponentially. This is what Einstein called the "cosmological constant." Alternatively, the dark energy might be associated with what physicists call a space-filling scalar quantum field that is not at its lowest possible energy. In that case, the dark energy would change with time, possibly causing the cosmic repulsion to decrease or even vanish. The way to determine whether the dark energy is a constant property of space or instead something that changes is to measure the history of the expansion of the universe and the growth of structure within it much more carefully than we have thus far been able to do, in order to see whether the dark energy has changed in the past. Several projects are under way to do this, possibly including a new satellite observatory.[6] JRP

7. With the introduction of "dark energy" to explain the expansion rate of the universe, are we throwing out the invariance of physical laws over time?

No, both possibilities described in the previous paragraph concerning the dark energy are in the context of standard modern physics: relativity and quantum theory. General Relativity allows a cosmological constant, as Einstein recognized. Quantum Field theory allows dynamical dark energy to be generated by a scalar field that is not in its lowest energy state; the dark energy would in that case be expected to decrease as the field "rolls" toward its ground state. Both possibilities assume that the underlying theory is unchanging. However, one of the goals of careful observations of our cosmic past is to check that assumption also. JRP

8. What's your reflection on the term *Double Dark* in regard to the more positive connotations of the word *light?*

The vast majority of the stuff in the universe seems to have no connection—or only the faintest connection—with light, so astronomers call it "dark." Part of the triumph of modern cosmology has been to discover the role of the very powerful but invisible aspects of the universe, dark matter and dark energy.

Dark matter is our friend. Dark matter creates the galaxies and all the other large structures that are held together by gravity. Without dark matter nothing else would exist. There would be no galaxies, no stars, no heavy elements, no rocky planets, and no life. So we owe a tremendous amount to dark matter. We scientists are hoping that if we can discover what the dark matter actually is, that could be an important clue to how the whole universe is actually put together.

Dark energy is even more of a tantalizing challenge because it's unclear how dark energy fits into our whole pattern of explanation in the physical sciences. If we can understand where it comes from, what its role is in the present universe, and how it might change in the future, then that might give us a very profound understanding. So although these things have very little to do with light, what we're hoping is that they're going to shed light on everything else.

On a more metaphorical level, does calling our universe the Double Dark Universe lend it a certain eerie, unpleasant connotation? Rather, think of it this way: we, the stardust part of the universe, are completely supported by these huge dark components without which we could not exist. This is the basis of what we are—that the light is supported by the dark. JRP

9. What were the main things that happened in the first hundred million years?

~ 10^{-32} seconds—Cosmic inflation creates small fluctuations and inflates them to much bigger sizes.

~ 10^{-30} seconds—After inflation ends, the universe fills with radiation and particles.

~ 10^{-10} seconds—Asymmetry between matter and antimatter develops, with about a single extra quark for every billion quarks and antiquarks, and a single extra electron for every billion electrons and antielectrons (positrons).

~ 10^{-4} seconds—Quarks and antiquarks annihilate, leaving a small remnant of quarks bound into protons, neutrons, and short-lived mesons.

~ 4 seconds—Electrons and positrons annihilate, leaving a small remnant of electrons.

~ 5 minutes—"Big Bang nucleosynthesis" creates the light nuclei and releases a lot of energy. Almost all the neutrons are bound into helium nuclei (formed from two protons and two neutrons), with a tiny fraction in heavy hydrogen (deuterium, with a proton and neutron) and light helium (two protons plus one neutron).

~ 400,000 years—Atoms form, and the universe becomes transparent to light. The cosmic background radiation starts on its way to us.

~ 100 million years—The fluctuations created during cosmic inflation grow into the first galaxy-size dark matter halos, and in them the first galaxies form. In these galaxies the massive first stars form and shine for about a million years before ejecting the first heavy elements. Their remnants are massive black holes, which become the first mini-quasars. JRP

10. How did quantum effects during cosmic inflation create small fluctuations in density from place to place?

The exponential expansion of the universe during cosmic inflation means that every point is surrounded by what astronomers call an event horizon, where space is moving away from that point at the speed of light. A black hole is also surrounded by an event horizon, and Stephen Hawking showed that General Relativity plus Quantum theory implies that whenever there is an event horizon there must be quantum fluctuations. The smaller the radius of the event horizon, the larger the quantum fluctuations—this is the origin of the "Hawking radiation" that hypothetical small black holes must radiate increasingly copiously as they decay. The quantum fluctuations occur in space-time itself, and that causes some regions to inflate a little more than other regions and thereby become less dense. Astrophysicists have derived the same results from several perspectives, so even though we don't yet have a full quantum theory of gravity, we are confident that we understand this aspect.[7] JRP

11. How do we know that for every surviving quark or electron, a billion quarks and electrons annihilated with their antiparticles?

Using the Standard Models of particle physics and cosmology, we can calculate that the amount of annihilation that must have occurred corresponds roughly to about a billion particle-antiparticle annihilations to one surviving particle (electron or quark).[8] The physics of the annihilation of electrons and their antiparticles (positrons) has been extensively studied both theoretically and experimentally, as have the annihilations of strongly interacting particles with their antiparticles. That's why we astrophysicists are confident that we understand the period of about a thousandth of a second to a few seconds, when such annihilations were occurring. The nuclear and atomic processes that occurred during the period from then to about four hundred thousand years have been studied even more thoroughly. JRP

12. Images of the large-scale structure of the cosmos reveal a filamentary cosmic web. Why does it look like that?

The explanation is a combination of the nature of the fluctuations in density that are created by cosmic inflation and the subsequent evolution of this increasingly inhomogeneous universe under the action of gravity.

The Russian astrophysicist Jacob Zel'dovich and his colleagues did an analysis about forty years ago that is the key to the explanation. Even before inflation theory, Zel'dovich and others had already intuited the sort of distribution of fluctuations on different scales that would be required to form the universe we see—today we call it the Zel'dovich fluctuation spectrum. Quantum effects during cosmic inflation naturally produce the Zel'dovich spectrum of fluctuations. Zel'dovich's analysis then identifies the likelihood of different kinds of collapse.

By far the most likely collapse is one-dimensional, which creates large, thin "pancakes" (Zel'dovich called them *blini* in Russian) of higher density. The superclusters of galaxies are such Zel'dovich pancakes; they have collapsed in one direction, but they are still expanding in the other two dimensions. The second kind of collapse that occurs is two-dimensional, along curved lines. Those are the first cosmic structures that reach reasonably high density, and that's what creates the basically linear or filamentary shapes that are so prominent in computer simulation visualizations such as figure 39. And the distribution of the galaxies really reflects that: on large scales it's these linear structures that dominate. Rarely, there is collapse in all three directions, and that leads to clusters of galaxies on big scales and individual galaxies on small scales. In visualizations like figure 39 of the dark matter distribution in the present-day universe, the dark matter halos of galaxies and small groups of galaxies are bright white pearls strung along the filaments, and the dark matter halos of galaxy clusters appear where big filaments cross. JRP

13. How do galaxies and galaxy clusters form?

Galaxies and galaxy clusters form in the expanding universe by a process called gravitational collapse. During the brief period of cosmic inflation at the beginning of the Big Bang, quantum fluctuations create slight differences in density from one region to another, and then these microscopic regions greatly inflate in size. They continue to expand with the expanding universe. Gravity makes regions that are slightly denser expand a bit slower—so as time goes on, they become denser than their surroundings. When a particular region is about twice as massive as average regions of its size, it stops expanding and "relaxes" by falling together, becoming what we call a dark matter halo. Meanwhile the rest of the universe keeps expanding and becoming less dense.

Gas cools and collects at the centers of the galaxy-size dark matter halos, and stars form from this gas. Nearby dark matter halos attract gravitationally and merge as time goes on, and sometimes the galaxies inside them collide and merge gravitationally, causing bursts of star formation and creating massive black holes. On larger scales, the merging of dark matter halos leads to the formation of groups and clusters of galaxies. This is a very complex process that cannot be calculated analytically, so in order to follow it we run supercomputer simulations. JRP

14. How do cosmological simulations work?

We start by using random number generators to simulate the quantum fluctuations in the early universe, and we evolve them as far as possible using simple mathematics. (This can be done on a small computer.) The results tell us how to set up the initial conditions for the simulation in a supercomputer: where to put each of the billions of particles that represent the dark matter, and what initial velocity to give each particle. We then let the particles interact with each other gravitationally—that is, the supercomputer calculates the gravitational force on each particle from all the particles. This changes the velocity of each particle, and the supercomputer then calculates where each particle will be a bit later. The forces on each particle are then calculated again, the particles are moved again, and so on. Fortunately, various methods have been discovered that speed up these calculations—although the steadily increasing power of supercomputers has been essential in enabling theorists to keep up with the rapidly improving astronomical observations, and often make correct predictions about the distribution and properties of galaxies even before the observations have been made!

The huge Bolshoi simulation in 2009 by Anatoly Klypin and Joel Primack took a little over two weeks of time using about fourteen thousand computer processor units (CPUs) on NASA's most powerful supercomputer, Pleiades, for a total of six million CPU-hours. (Running this computation actually took about two months, since we had to do test runs and the big simulation while the supercomputer was being built and tested and because parts of the computation had to be rerun several times. Many more months have been required to analyze it.) The Bolshoi simulation is currently the highest-resolution large cosmological simulation, and it was run using the latest cosmological parameters.[9]

We stored the locations and velocities of all the eight billion particles at many time steps during the Bolshoi simulation, roughly every fifty to one hundred million years of simulated time. We then had the

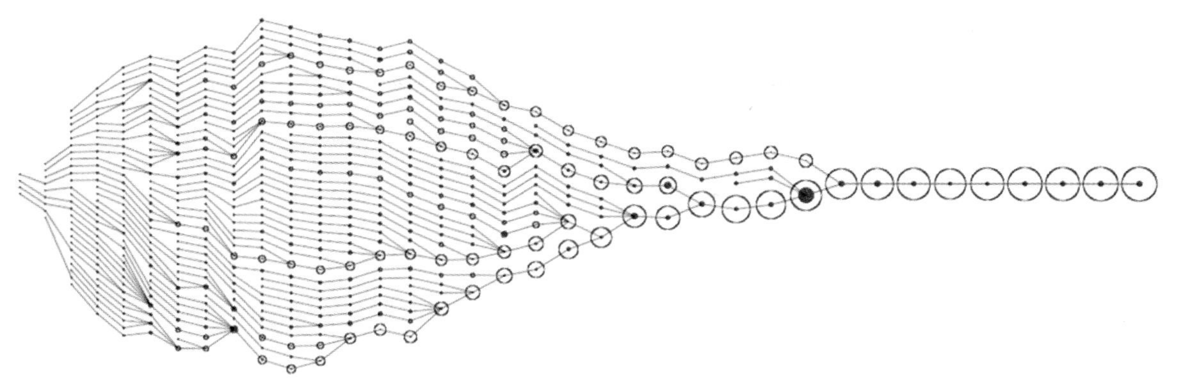

Fig. 72. The merging history of a large dark matter halo

computer find all the gravitationally bound dark matter halos, including halos within other halos, at every stored time step—about ten million halos at any given time, about fifty million halos over the course of the simulation. Finally, we had the computer determine how the halos at a given time step merged to form the halos at subsequent time steps. This allowed us to construct the complete "merger tree" of the simulation. Primack's former graduate student Professor Risa Wechsler and her group at Stanford University finished this part of the computation. The merger tree of a single large halo is represented by figure 72.

The merger tree is the basis for what we call semi-analytic models (SAMs) that allow theorists to follow the evolution of millions of galaxies as they form and merge to create the entire population of galaxies in a large region of the universe. These SAMs include all the processes that are thought to be important in galaxy formation and evolution. Several groups of theorists are running several different SAMs, which model the galaxy formation process differently. As we compare the resulting predictions with observations of both nearby and distant galaxies, we will see which of the models is more accurate—and we can incorporate that information to allow us to do better calculations in the future and to understand the processes that form galaxies, including our own Milky Way. JRP

15. How do you simulate the formation of a single galaxy?

We have to treat dark matter and ordinary (atomic) matter separately. For the dark matter, only gravity is important, while for atomic matter the other forces also have to be taken into account. These forces allow clouds of atomic matter to heat when they collide, and then to lose energy and cool by emitting radiation. When the gas cools and becomes dense enough to form stars, the energy that the stars radiate, including those that become supernovas, can heat nearby gas enough to expel some of it from the forming galaxies. The stars and supernovas also enrich the forming galaxies with heavy elements, which makes it easier for the gas to cool and for stars to form, now along with planets.

Sometimes galaxies collide. When this happens, much of the gas in the galaxies flows to their centers, where great bursts of star formation occur. The galaxy centers merge and coalesce rather quickly. Only about a thousandth of the gas is accreted by the supermassive black holes at the galaxy centers, but the energy that this releases is comparable to all the energy released by the stars during their entire lifetimes. Indeed, when the gas accretion rate peaks, the black hole becomes a quasar that greatly outshines all the stars in the galaxy, because it is releasing a large fraction of the $E = mc^2$ energy in the mass of the infalling gas.

The computer codes to treat all of this complex astrophysics are of two basic types, called smooth particle hydrodynamics (SPH) and adaptive mesh refinement (AMR). SPH codes treat the gas as particles, with each particle having the average properties of the gas in its vicinity, while AMR codes divide the space up into cells that are more closely packed where the gas is dense or rapidly changing. SPH codes are fast but crude; AMR codes are about ten times slower but more accurate, especially at capturing such complex phenomena as the shocks that occur when gas clouds collide.

When lots of stars form, those that are eight or more times as massive as the sun shine a thousand or more times more brightly than the sun, and after a few million years they have exhausted the hydrogen in their centers. They then become huge red giants, and in their centers they are fusing larger nuclei to make still heavier nuclei. Soon they run out of nuclear fuel, and their centers collapse into neutron stars or black holes, ejecting the outer parts of the star as a supernova. The heavy elements become stardust, which permeates and surrounds the star-forming region and absorbs much of the starlight, emitting it as longer-wavelength heat radiation. Primack's former graduate student Patrik Jonsson, now at Harvard University, developed a computer program called *Sunrise* to simulate all these processes and produce realistic images of forming and interacting galaxies such as figure 49. The *Sunrise* code is now widely used by other galaxy simulators around the world.

The capabilities of present-day supercomputers are not yet adequate to simulate more than a few galaxies with resolution sufficient to include the main astrophysical processes that create galaxies. Therefore, we use high-resolution galaxy simulations to determine the key properties of forming galaxies of various masses, and then we use semi-analytic models (SAMs) to follow the evolution of the entire galaxy population. SAMs predict the numbers of galaxies of various sizes, luminosities at many wavelengths, and various other properties, including the numbers of mergers and quasars as a function of cosmic time, as well as the numbers of groups and clusters of galaxies. Comparing their predictions with our improving observations tests these simulations. Many aspects are correctly predicted, and those that are not challenge us to improve our simulations and our understanding. JRP

16. What is the role of computer simulations in modern astronomy?

Traditionally, astronomers were either observers or theorists. But as our physical understanding of astrophysical processes has improved and as computers have gained in processing power, a distinct subspecialty has developed to run and analyze simulations. It is not wise, however, to treat computer simulations as a "black box" that magically produces answers. Instead, the beauty of computer simulations is that they not only let us treat extremely complex processes but let us vary the underlying assumptions and study intermediate steps, so that we can understand why and how the simulations produce their results. We often find it useful to invent simplified "toy models" that capture the essential processes that the full-scale simulations treat in detail. Visualizations of supercomputer simulations are increasingly important to help scientists understand the computations and their results. Such visualizations are often beautiful, and they are helpful in communicating the science to students and the broader public. JRP

17. What are the most common elements in us and in the universe?

The top five elements in the sun and most other stars are hydrogen, helium, oxygen, carbon, and nitrogen. The proteins out of which our bodies are made are composed almost entirely of carbon, hydrogen, oxygen, and nitrogen (CHON for short)—the same elements that are most abundant in the sun except for helium. (Helium plays no role in our bodies since it is chemically inactive.) Proteins are made of twenty amino acids, and only two of these are made of any other element besides CHON (methionine and cysteine also contain sulfur; selenocysteine, which contains selenium, is sometimes included in a list of twenty-one amino acids). The nucleic acids that store genetic information are also made of CHON, plus phosphorus. The human body is about 61 percent oxygen (mostly in water molecules), 23 percent carbon, 10 percent hydrogen, 2.6 percent nitrogen, 1.4 percent calcium, and 1.1 percent phosphorus. The last percent or so is other kinds of atoms. JRP

18. If we say that the earth, the sun, or a galaxy is moving, what is it moving with respect to?

There is a preferred reference frame at every point in space and time, in which the cosmic background radiation has the same temperature in all directions (except for tiny fluctuations reflecting the small differences in density from region to region in the Big Bang). In Special Relativity, which applies to motion in straight lines at constant speed in empty space, there cannot be any special reference frame. But our universe isn't empty space—it is filled with the cosmic background radiation and the light from all the galaxies, among other things. If one is moving with respect to the reference frame of the cosmic background radiation, the radiation will appear blueshifted in the spaceship's front window and redshifted looking out the rear. Measurements of the cosmic background radiation using NASA's Cosmic Background Explorer (COBE) satellite showed just these sorts of redshifts and blueshifts as the earth moved around the sun, and these are now being measured even more precisely by NASA's Wilkinson Microwave Anisotropy Probe (WMAP) satellite and the European Space Agency's Planck satellite. But these measurements were not the first proof that Copernicus was right that the earth orbits the sun. The British astronomer James Bradley discovered in 1725 slight seasonal changes in the apparent direction of stars caused by the earth's motion around the sun, a phenomenon known as stellar aberration. JRP

19. What is the average speed at which galaxies are moving apart from each other? How far into the future will it be when galaxies are moving apart at light speed or greater?

Since the speed of a galaxy away from us is proportional to its distance in a uniformly expanding universe (Hubble's Law), it wouldn't make sense to quote an average speed. Instead we can say that galaxies at a distance of 100 megaparsecs are moving away at 7,000 h_{70} kilometers per second, that those twice as far away are moving twice as fast, and so on. (One megaparsec = 1 Mpc = 3.26 million light-years, and according to the latest observations $h_{70} = 1.00 \pm 0.02$.)

Since we don't yet know what the dark energy is, we can't say for sure what will happen in the distant future. But if the dark energy is

Einstein's cosmological constant, then the nearest cluster of galaxies, the Virgo Cluster, which is now about 60 million light-years away from us, will be moving away from our galaxy at the speed of light in about 100 billion years. JRP

20. How is it possible for distant galaxies to move away from us faster than the speed of light? Doesn't this violate Einstein's relativity?

Special Relativity says that the speed of light (in a vacuum) is the ultimate speed limit in "inertial reference frames" (moving with uniform velocity and with no gravity). But in General Relativity (which includes gravity), inertial reference frames are dragged apart by the expansion of the universe. The velocity with which distant inertial reference frames are dragged away from us by the expansion of the universe is proportional to distance (Hubble's Law), so for a large enough distance the velocity exceeds the velocity of light. This is all according to Einstein's General Theory of Relativity. JRP

21. How can our cosmic horizon represent the present location of distant galaxies as they move away when the light by which we see them is coming from the past? How far away are they now?

When we see a ship on the horizon, its actual distance is essentially the same as its apparent distance, since the time it takes the light to get to us from the ship is negligible. But when we see a galaxy across the universe, the light travel time can be nearly the age of the universe. Cosmologists define the (particle) horizon to be the sphere surrounding us that represents the present locations of the most distant objects that could have been detected by us via signals that they emitted at earlier times. The question is about the distance to the horizon. Of course, the distant galaxies have moved away from us since they emitted the light that we see now. But once we know the basic cosmic parameters—how much matter and dark energy, and how fast the universe is expanding now—we can use General Relativity to calculate where each galaxy is now (that is, when its cosmic clock, started at the Big Bang, reads the same as our cosmic clock). The distance to the matter that radiated the heat radiation of the Big Bang that we are receiving now is about 46 billion light-years. The distance to the cosmic horizon is only slightly greater than that, about 47 billion light-years. How can this matter, which long ago turned into galaxies, be so far away, when the time since the Big Bang is about 13.7 billion years? The answer is that these galaxies have for some time been moving away from us faster than the speed of light. In fact, the universe has expanded by about a factor of a thousand since the heat radiation of the Big Bang started on its way to us. JRP

22. Is there still a chance that the universe is going to end in a Big Crunch?

All the modern evidence points to exactly the opposite future, a universe not collapsing in a Big Crunch but rather expanding indefinitely on the large scale. But since we don't yet understand the nature of the dark energy that's making the universe expand faster or the fundamental nature of gravity, we have to keep an open mind about the ultimate endpoint. JRP

23. What's beyond the visible universe?

There is every reason to think that the universe beyond our cosmic horizon is just like the universe nearby us. But on really gigantic scales, the theory of Eternal Cosmic Inflation suggests that there are other universes (or "cosmic bubbles") that we are permanently cut off from, as we've tried to explain in chapter 7. It is possible that the laws of physics are entirely different in these other universes. JRP

24. Why would different laws of physics in other universes not let life exist?

It is possible that the laws of physics that control things like chemistry are actually determined by the processes that occur at the end of cosmic inflation, so the laws could be different in different cosmic bubbles than ours. Physicists have had fun considering the effects of many different changes in the laws of physics, and they have found that a universe where the laws of physics were even slightly different might be inhospitable to life as we know it. For example, stars might not exist, or they might have short lifetimes. The element carbon, the basis for the organic chemistry that underlies all living organisms on earth, might not be formed in stars.[10] JRP

25. Why is there a smallest size?

General Relativity tells us that there's a maximum amount of mass that can be squeezed into a region of any given size. If more mass is packed in than the region can hold, gravity there becomes too intense, and the region itself—the space—collapses to no size at all. This is a black hole. Meanwhile Quantum Mechanics sets the minimum size limit, but in a very peculiar way. Electrons, protons, and other particles have extremely small masses and are always whizzing about. They are very hard to pinpoint. The "size" of a particle is actually the size of the region in which you can confidently locate it. The smaller the region in which the particle is confined, the more energy it takes to find it, and more energy is equivalent to larger mass. There turns out to be a special, very small size where the *maximum* mass that relativity allows to be crammed in without the region collapsing into a black hole is also the *minimum* mass that quantum mechanics allows to be confined in so tiny a region. That size, about 10^{-33} cm, is called the Planck length, and it's the smallest possible size. In string theory, sizes smaller than the Planck length get remapped into sizes larger than the Planck length. JRP

26. Does the Cosmic Uroboros in some ways fulfill Einstein's hope of finding a unified theory?

Partly. In fact, the tail-swallowing represents the hope that we'll finish the job and connect the forces that are most important on small scales with gravity, which dominates on large scales. What physicists now call the Standard Model of particle physics—which was the work of many scientists twenty to thirty years ago—achieved a partial success in unifying the theories of electromagnetism and the weak and strong forces. We now know that those theories have a very similar structure, which we call gauge theory. Basically, this means that the forces that control on the scale of the atomic nucleus are analogues of electromagnetism. The challenge that remains is to unify all of fundamental physics including gravity and in particular to understand the nature of the main components of the universe—dark matter and dark energy. It's possible that superstring theory will be part of the answer.

We're hoping that major new discoveries will come in the next few years. There's a good chance that if the dark matter is supersymmetric Weakly Interacting Massive Particles (WIMPs), then we're going to discover it in the next few years—and maybe these new discoveries will give us the clues that Einstein didn't have, and that we physicists today still don't have, that may help us unify the whole picture. JRP

27. Could you explain string theory and its possible connection with the Cosmic Uroboros?

String theory includes a quantum theory of gravity, along with having the possibility of including all the other fundamental forces of nature that we know about—the strong, weak, and electromagnetic forces. So it's a possible unified theory. And in the mathematical structure of the theory, if you try to go smaller than the Planck length, you can't; what you think is smaller turns out to be bigger. So string theory automatically embodies the idea that there's a smallest size: the Planck length.

The trouble with superstring theory is that it's unfinished. It's a theory of ten or eleven dimensions (depending on how you think about it), whereas we can probe, with our experiments and observations, only the three dimensions of space and one of time. We don't know how to make string theory predictions about our three-plus-one dimensional world. There appear to be a very large—possibly almost infinite—number of ways, and we don't know which (if any) makes sense! That's the "compactification" problem of string theory. And until we have a better understanding, or a better theory, we're not yet able to turn this wonderful mathematical entity into a scientifically predictive theory.[11]

I tell my students that string theory is probably a project for middle-aged physicists. It's a very dangerous subject for young physicists because it's so difficult, and it's all too likely that if you start down that path you're not going to end up with a job. It's also not well suited for old physicists like me, because I'd like to find out whether theories I work on are true or not, and it doesn't seem likely that we're going to find that out about string theory anytime soon. So I think string theory is a job for middle-aged physicists—with tenure. JRP

28. Does the Double Dark theory help explain the Big Bang or what happened before it?

Our best understanding of the Big Bang is that cosmic inflation—an extremely brief burst of exponential growth of the universe—started off the expanding universe and provided the initial conditions for the formation and evolution of galaxies, galaxy clusters, and superclusters according to the Double Dark theory. From the first paper on Cold Dark Matter,[12] cosmic inflation was part of the theory. But the Double Dark theory doesn't explain cosmic inflation. As we discuss in chapter 7, the best idea we now have for the origin of cosmic inflation is that it is a local transition from "eternal inflation" (which continues in most of the universe on the largest scales) to the expanding universe. JRP

29. How is the cosmic inflation–human inflation analogy encouraging?

What we're trying to emphasize here is that, like the newborn universe after cosmic inflation, humanity is very young with a possibly immense future. Our species is perhaps a hundred thousand years old. Civilization—let's say the Neolithic period, with the beginnings of cities and division of labor—started around ten thousand years ago, at the end of the last ice age. And the Industrial Revolution, with the concomitant exponential increase of human impact on our planet, began only about two hundred years ago. But the possible future for humanity and our descendants is measured in hundreds of millions of years at least, if not billions. And there will be stars shining in the Milky Way for trillions of years. In fact, since small stars get brighter with time, the Milky Way will be brighter a trillion years from now than it is today. So the relative time between the evolution of modern humans and the potential future for our descendants is truly enormous—by factors of millions. And that's not so different from this tremendous difference between the very brief period of cosmic inflation and the very long period of slow expansion and cosmic evolution that has followed inflation. So the analogy that we want to emphasize is that we've got to end the present brief period of human inflation in some sensible way, so that there's a nice environment for many millions or billions of years into the future. There is no law of physics that says we have to fail. JRP

30. Why can't there be an earth-size or even a galaxy-size consciousness?

Connections through electronics—the Internet, for example—give the possibility that the whole earth can in some sense become a thinking creature. But if you look at how thinking actually works, even in the human brain it seems that there are a lot of separate parts in which the complex operations that underlie thinking are taking place. Certainly in large supercomputers, the really hard calculations are being done in individual processors, and what slows it all down is the communication between these processors, which is ultimately limited by the speed of light. So one could have a society that exists on much larger scales, but the individual consciousness—where ideas form and travel quickly—will have to be much smaller. A galactic-scale consciousness, for example, will be very slow because it takes a hundred thousand years for light to cross our galaxy. The number of galactic thoughts in the whole history of the universe will be quite limited compared to the number of thoughts that you can have in a day. JRP

31. Can't there be near instantaneous communication using quantum nonlocality?

Quantum nonlocality is an important phenomenon with implications for quantum computing and other future technologies but not for faster-than-light communication.[13] JRP

32. What's wrong with dualistic thinking?

Plato, in his famous myth of the cave in *The Republic,* argued that the world that we see around us is an illusion—that what is real, or at any rate *most* real, are the forms, and that in the world of material objects there exist only poor reflections of these ideal forms. An example of a Platonic form is a circle—and the circle that you can draw on a blackboard is indeed a poor reflection of the perfect circle, the ideal circle.

Modern physics has completely changed our view of this: as far as we can tell, the structures of atoms and all small things like that are perfect mathematical forms. And yet that's what matter is! So this apparent distinction between matter and form has completely disappeared in modern physics. I think that undermines not only Platonic but also Cartesian dualism.

Cartesian dualism, which emphasizes the difference between mind and matter and between reason and emotion, is very misguided because it implies that there's a fundamental difference between spirit (or "ideal forms") on the one hand and the gross material universe on the other hand.[14] That's a misunderstanding of how the world works. It basically represents an unwarranted extrapolation of the physical world that we have in our minds (things that are solid and obviously here) beyond the familiar scale—namely, things that are about as small as we can see up to things that are about as big as we can see. But with the help of instruments, we can see much smaller and much larger things—and they don't behave the same way. To understand how the world works, you have to understand how things behave on different scales. JRP

33. Since you are certainly not advocating the pre-Copernican vision of the universe, are you saying we're the center of the universe in an ironic way?

From the Greek era all the way through the Middle Ages until the Renaissance, people thought that physically Earth is literally at the center and that the whole universe revolves around Earth every day. We now know that isn't true. But now we also know that we are at the center of our own observable universe and that as we look out in space we look back in time. This applies to every location in the universe. We have also explained that we intelligent creatures also have a central or special place in the modern universe in several other ways. We are not using this "center of the universe" language ironically, but there is perhaps some irony in the fact that after centuries of believing that science has pushed us out of the center of the universe, we discover that we're central after all. JRP

34. What does cosmology have to do with morality?

The basic moral imperative is survival. That has been the essence of all life. All of us are the result of an unbroken chain of survival and procreation—so that has to be built into our world picture. The most important insight that I think we learn from cosmology is that the universe is very old, that we humans just arrived on the scene, and that our descendants can enjoy an indefinitely long future. The story from Genesis of a very short-lived universe that we humans have lived in for all but the first five days is completely wrong. If you understand instead that we're living in the middle of time, with a great deal of time ahead of us if we don't waste our opportunities, and you combine that with a basic desire for survival, then you should be motivated to help make the changes necessary to make a graceful transition from exponentially increasing human impacts to a sustainable relationship with our home planet. JRP

35. The economy works because of an expectation of future growth. If the economy stopped growing, it would collapse, and that would lead to environmental collapse. How do we avoid that?

The economy can grow, but at a sustainable rate. Up until 1974, when the first oil shock occurred, it was more or less an axiom of economics that energy use grows in lockstep with the general economy. But since 1974, economic growth in the United States has been much larger than the growth of electricity or energy use. And in California, electricity use per capita has barely grown at all since 1974—per capita energy use in California is now half that of the United States as a whole. How is it possible that California has done so well? Because a political decision was made: the California Energy Commission was created, and there was a major effort by physicists (especially at Lawrence Berkeley National Laboratory) to develop ways of measuring energy use by appliances. This resulted in new regulations, and soon half the refrigerators made in America could not be sold in California because they were so profligate in energy use. (Because of cheap energy, manufacturers stopped putting insulation into refrigerators. Water would condense on the cold exteriors, so electric heaters were put in the walls and doors of refrigerators!) Refrigerators today are twice as big but use less electricity than the refrigerators that were the average back then. Because California is a big part of the economy, the whole country has followed suit. The same thing happened with heat pumps and with general building energy use.[15]

The secretary of energy, Steve Chu, a Nobel laureate physicist who was previously director of Lawrence Berkeley Laboratory in California, is now advertising these facts and implementing them at the national level. So we can break these habits of greatly increasing use of resources without wrecking the economy. The US economy is largely a service economy and, to the extent that we sell things, a lot of what we sell is information—which doesn't require a lot of resource use. So I think that creativity can really make up for a lot of resource use.

And we don't have to stop growth, we just have to stop the exponential growth in resource use—so it's not as great a challenge as people often imagine. A change in direction in the near term can have enormous long-term consequences. JRP

One of the key reasons we feel cosmic inflation is the right model for how to end the inflationary growth of resource use is that it does not require ending growth. Growth continues, but a lot more slowly. The transition will not be gradual or easy everywhere, although it can be gradual and easy in some places. There will certainly be a lot of suffering, no question. But there's going to be a lot more suffering if people try to keep going as they are.

Joel and I are trying to take a big view. Rather than explaining how to change the economy or how to finesse this or that detail, what we're trying to do is step back and ask, "What do we all share? How could we possibly have a set of principles that could guide us so that those people who are experts in their respective fields, if they bought into these principles, would start thinking in a new way?" I personally don't know how to restructure the economy. But if people who understood the economy also understood these principles that come out of cosmology as analogies and perhaps as models, they would be empowered to rethink the economy in very interesting ways and in new terms. It's going to take much more than a village; it's going to take a planet to do this, and that's why we need people in all countries, especially the big developing countries including China, India, and Brazil, to be part of the kind of cooperation that could transform the human world.

We have to start, though, with something we all can agree on. No one ever starts by agreeing on details. People can very often start by agreeing on basic principles before they see how those principles are actually going to affect the details: that's how you get the first agreement. Then, once they've committed themselves to those principles, they have to translate those principles into action, and that's where the nitty-gritty work enters. Let's fight it out politically—fine. But we have to begin with broad agreement on something much bigger, and that's why we start with cosmological principles. NEA

36. What do you mean by "myth," and where did this idea of a modern myth actually come from?

The word *myth* is often used to mean a false belief, somebody else's notion that's just simply wrong, or maybe a quaint old story. But a cultural myth is an explanation of the larger reality *that you're a part of.* So naturally if it's someone else's myth, it doesn't feel true to you, but if it's your myth—if it explains your origin and the meaning of your position in your society—then sometimes you don't even realize that it exists. You're like a fish unaware of the water it swims in. Every traditional culture has a shared sense of what the world beyond the ordinary material world is like and what it means.

For the idea that the modern world still needs myth, we (and many others) owe a tremendous debt to two thinkers in particular: Mircea Eliade, a Romanian historian and philosopher of religions at the University of Chicago, who inspired me when I was a student there, and Thomas Berry, a Catholic priest, cultural historian, and, in the term he coined for himself, "geologian." Eliade opened many people's minds to the realization that myths, which were commonly assumed to be folkloric tales mainly of obsolete gods, are actually humanity's way of looking through the ordinary world to the meaning within and experiencing the existence of something sacred. The metaphor system that any given culture uses to express its sense of the sacred is local to that culture, but the experience is universal. (Even those who dislike the religious tone of the word *sacred* almost always hold something sacred—for example, the scientific method or freedom or truth or, in the case of the signers of the Declaration of Independence, their honor.) Eliade worried that without the kind of experience of sacredness communicated by myth and entered through myth, we modern people might not really value our civilization enough to maintain it.

But simply learning that your society needs a myth does not give you one, and no prescientific myth in an age of science can be believable across cultures. It was Thomas Berry who showed that we actually have the basis for a myth that's huge—that explains the whole earth. He didn't know about modern cosmology (the New Universe picture hadn't been fully developed at the time he wrote his pivotal book, *The Dream of the Earth*), but he mythologized the earth. He gave us the first mythology of the whole earth that all people could be part of. And that was truly an enormous contribution because until then there had been local mythologies but nothing everybody *could* share if they simply understood it. Later Berry collaborated with Brian Swimme on the popular book *The Universe Story,* which is perhaps the first attempt to mythologize the modern universe. NEA

37. Isn't metaphor more relevant to literature than to science?

It's not a choice of hard science versus soft metaphor. Science is meaningless without metaphor, especially science that pertains to those aspects of the universe that we humans have no direct contact with. There would be nothing but equations and unexplained drawings; scientists wouldn't be able to talk to each other to try to figure out what their equations mean without metaphor.

Human beings always think in metaphor about everything abstract. If you try to think of anything abstract, you must compare it to something else that's more concrete. In fact, even *concrete* is a metaphor. We cannot escape that; that's how our brains work.[16] Understanding the universe is a combination of understanding the "hard science" and understanding how we humans think. If you don't take into account the neuroscience of how we think—as well as the history of our religions as to how we have behaved—then you're not going to be able even to explain the hard science. We're trying to build a bridge from what's considered "hard science" to the rest of culture. We ask the question, "Having realized that we live in a universe different from what anybody assumed, what does this mean for us?" In order to answer that question, we absolutely must use metaphor.

But we can't just choose a favorite religion and try to use the metaphors of that religion to explain the universe—for example, stretching the six days of creation at the beginning of Genesis to make it seem as though it actually explains the Big Bang. That's ridiculous—it doesn't do justice to the Bible or to science. You can't choose your metaphors first and try to squeeze science into them. But once you understand the science, you have an obligation to try to find metaphors that work. NEA

38. Where did the idea of Scientific Mediation come from, and if it's so effective, why isn't it being used?

The story of Scientific Mediation began in the 1970s, when I was working for the Ford Foundation as a lawyer, researching alternative means of dispute resolution, other than courts, that Ford might fund. I was invited to Washington to a meeting of a subcommittee of President Gerald Ford's Science Advisory Committee, and met there a group of very high level scientists focusing on a huge problem: that it's extremely hard to get good science into policy making, but failing to do so can lead to disaster on multiple size scales. The subcommittee was trying to figure this out: how can you get good science advice into government agencies when the officials with the power to decide understand the underlying science only minimally and are far more responsive to the economic and political arguments? The problem has not been solved to this day. Suppose you are such an official. You're under time pressure to make a policy decision. Your favorite science advisers disagree. How are you supposed to know what's true if they don't? If you appoint an impressive committee of experts and demand that they reach a consensus, they will have to paper over their differences, which might not help you reach a wise decision. If you don't demand a consensus, however, they will give you a majority and a minority report, which also might not help you since you know that scientific truth is not determined by majority rule. As Albert Einstein said when told that a hundred Nazi professors disputed his Jewish Theory of Relativity, "If the theory were wrong, one professor would be enough."

The subcommittee was considering a proposed solution: the agencies should subtract the scientific part of the controversy from their deliberations and refer those questions to a "Science Court." In the Science Court a panel of scientists would judge while two expert but disagreeing scientists would act as advocates and present the two sides of the case. The panel of judges would then determine the scientific truth for purposes of the decision at hand. From the point of view of the government agency, the Science Court would be like a black box, and answers would be handed out.

I listened to this discussion in bewilderment; it involved a misunderstanding of both courts and scientific method. The ultimate purpose of courts has never been to find out the truth but to resolve disputes in a fair way without violence. In a court, truth is what's found believable by a judge or jury following accepted procedures. So courts were the wrong model when the purpose was to get as close to scientific truth as possible at the moment. Science has its own way of proceeding, and there's no easier, quicker route to knowledge than carefully pursuing the scientific method. Nevertheless, political decisions do have to be made, and when there is no time to wait for more research, the important thing is to make clear the range of possibilities that are open because the science is not complete, and to help people understand *why* different expert advisers will come down in different places along that range. Having worked at the Ford Foundation on mediation in nontraditional contexts, I realized that by using mediation techniques but respecting the method by which science makes progress, it might be possible to solve the subcommittee's problem and produce a report that not only reflects the state of the art of knowledge but changes the debate. The goal is to illuminate the situation, including the stakes if we are wrong, without pretending that more is known than actually is.

Scientific Mediation is a procedure whose results are difficult to manipulate, and this is why it's a dangerous approach for a government agency that already knows what it wants to do and is looking for a respected group of scientists to give it a stamp of approval. Unexpected but very valuable truths are likely to emerge from Scientific Mediation about why certain economic and political biases tend to push a scientist one way or the other in his or her interpretation even of scientific data. These are things that the agency might not want to know or want others to find out, and in a highly politicized context it would take courage and a genuine commitment to the public interest to try it. This may be why no one has done so in the United States.

In the nuclear waste controversy described in chapter 6, however, the Swedes were looking for the truth; they really wanted to know if their nuclear waste disposal plan was going to work or not, and I deeply admired that. In fact the Scientific Mediation that we did there revealed problems with their nuclear waste disposal plan that forty-three other studies did not discover. Discovering these and other problems led Sweden to perform further iterations of the nuclear waste disposal plan, which is precisely the wise course. When the science is against you, it's crazy to deny it—or to claim that it's not well-enough established to worry about, as so often happens here in the US. The right way to proceed is to accept the new knowledge graciously and improve your plan. NEA

39. The multiverse can't be proved by science, so that makes it a faith-based system. If so, why should we adopt it?

No one should "adopt" or "believe in" the multiverse or eternal inflation. At this moment in the history of science, there is a clear line between physics and metaphysics, and that line is represented by the instant of cosmic inflation. The theory of Cosmic Inflation has made actual predictions, and all the predictions that have been tested have been confirmed by observation. Therefore the theory of Cosmic Inflation is science. But what happened before Cosmic Inflation? There's no evidence that can yet allow us to answer this question. We're not saying you should have faith in eternal inflation; we're saying you should have faith in the scientific method—and the possibility that we may someday be able to test Eternal Inflation theories. If we do figure out how to do so and we find out that Eternal Inflation is wrong, it's wrong! So this is not something to have faith in; this is something to call metaphysics today and be open to the possibility that some day it may be physics. NEA

40. What role does progressive/enlightened religion have in your idea of a cosmic society? Do you value religion at all?

We value it very highly. Some of the greatest and most important discoveries made about how human beings can relate to each other, how we can love each other, how we can bond, how we can build communities and civilizations, how we can live lives of principle above ego, and how we can feel ourselves connected to the larger, invisible universe, however we may conceive of that universe—all these things were discovered and developed by religions. Every religion has created a metaphorical language to bring its members into connection not just with other people but with the invisible—with the meaning of their universe. We modern people still need all that. We need to borrow many ideas and images from religions so that *we* can understand how to connect to our universe—the one that we now know exists. But in every religion there are also mistakes, which have developed over time as different people's experiences, beliefs, and opinions got embodied in the history of the religion. Some of these mistakes are very serious indeed and potentially fatal. And we hope, as Dwight Terry said in his bequest that funded the lectures this book is based on, that assimilating science into culture will actually lead to a "more broadened and purified religion," by which we hope he meant purified of mistakes—purified of the parts that don't serve us as human beings. A religion purified this way could guide a broader range of people through the uncharted waters we are entering.

The idea that the cosmic society can be an Ark is very important. It doesn't mean that we believe Noah saved two of every animal (or seven of the clean ones). We don't need to take these stories literally, but we need to look for the meaning of stories that speak across the ages and ask each other, "What can we learn from this?" Because the fact is, religious imagery has more resonance than anything else. You can't invent such a potent language; a language has to arise from human experience. If we don't use religious imagery, then our pedestrian descriptions of the new universe are probably going to be as successful as Esperanto. No one can invent a whole new series of images to connect us to the universe—because we won't feel anything. We need to build on the best of our cultures. If religions hold tightly to literal understandings of scriptures, they cannot serve the cause of world peace or the spiritual needs of an emerging global civilization. But if enlightened religions are willing and able to expand to encompass new knowledge, then there is a tremendous role for them, and they can be an irreplaceable part of the long-term solution. NEA

Notes

Chapter 1. The New Universe

1. See FAQ-1.

2. "I like to summarize what I regard as the pedestal-smashing messages of Darwin's revolution in the following statement, which might be chanted several times a day, like a Hare Krishna mantra, to encourage penetration into the soul: Humans are not the end result of predictable evolutionary progress, but rather a fortuitous cosmic afterthought, a tiny little twig on the enormously arborescent bush of life, which, if replanted from seed, would almost surely not grow this twig again, or perhaps any twig with any property that we would care to call consciousness." Stephen Jay Gould, *Dinosaur in a Haystack: Reflections in Natural History* (New York: Three Rivers Press, 1995), 327.

Chapter 2. Size Is Destiny

1. On cosmology and morality, see FAQ-34.

2. That the smallest size is the Planck size is explained in FAQ-25.

3. The significance of the tail-swallowing is discussed in FAQ-26, and its possible connection with String Theory is explained in FAQ-27.

4. Why much larger consciousness is impossible is explained in FAQ-30.

Chapter 3. We Are Stardust

1. For more on what elements we are made of, see FAQ-17.

2. For more on what the dark matter might be, see FAQ-4. Neutrinos (not shown on the Cosmic Density Pyramid) make up between 0.1 percent and 1 percent of the cosmic density, as explained in FAQ-5. For more on the dark energy, see FAQ-6 and FAQ-7. We can include both mass and energy density in the Cosmic Density Pyramid using $E = mc^2$, as explained in FAQ-3.

3. See FAQ-1.

4. For more on how galaxies form, see FAQ-13 and FAQ-15. For more on supercomputer simulations, see FAQ-14 and FAQ-16.

5. For an explanation of why the cosmic web of dark matter in the Bolshoi simulation looks so filamentary, see FAQ-12 and FAQ-13.

Chapter 5. This Cosmically Pivotal Moment

1. "And some of the great images of the Apocalypse move us to strange depths, and to a strange wild fluttering of freedom: of true freedom, really, and escape to *somewhere,* not an escape to nowhere. An escape from the tight little cage of our universe; tight, in spite of all the astronomists' vast and unthinkable stretches of space; tight, because it is only a continuous extension, a dreary on and on, without any meaning: an escape from this into the vital Cosmos, to a sun who has a great wild life, and who looks back on us for strength or withering, marvelous, as he goes his way." D. H. Lawrence, *Apocalypse and the Writings on Revelation,* The Cambridge Edition of the Works of D. H. Lawrence, vol. 2, ed. Mara Kalnins (Cambridge: Cambridge University Press, 1980), 76.

2. Like geologists and evolutionary biologists, astronomers reconstruct the past to understand the present. Landforms erode and only a tiny fraction of organisms fossilize, but all the energy that was ever radiated by galaxies is still streaming through the universe and can be detected in some form. Some of this radiation is altered. For example, redshifting occurs because the wavelengths of photons stretch as the universe continues to expand, and some short-wavelength photons like X-rays and ultraviolet light are absorbed by dust and re-emitted at longer wavelengths. To figure out what happened in the cosmic past, we must see the entire electromagnetic spectrum, from the high-energy gamma rays to the long-wavelength radio waves. Fortunately, NASA's Great Observatories in space cover much of this wavelength range, from short to long wavelength: X-rays (the Chandra X-Ray Observatory), near ultraviolet and visible light to the near infrared (the refurbished Hubble Space Telescope), and infrared (the Spitzer Space Telescope). These great spacecraft observatories have been joined by the European Space Agency's XMM Newton X-ray

telescope and Herschel far-infrared telescope. For more on this, see Joel Primack, "Hidden Growth of Supermassive Black Holes in Galaxy Mergers," *Science,* 30 April 2010, 576–578.

3. This is discussed in D. G. Korycansky, G. Laughlin, and F. C. Adams, "Astronomical Engineering: A Strategy for Modifying Planetary Orbits," *Astrophysics and Space Science* 275 (2001): 349–366. The article explains in detail how planetary orbits can be changed by altering the orbits of large comets. It concludes by saying: "An obvious drawback to this proposed scheme is that it is extremely risky and hence sufficient safeguards must be implemented. The collision of a 100-km diameter object with the Earth at cosmic velocity would sterilize the biosphere most effectively, at least to the level of bacteria. This danger cannot be overemphasized." If our distant descendants use this method to keep Earth's climate balmy despite the increasing strength of the sun's radiation, they can design the orbit of the comet so that the net momentum imparted to the earth is in the right direction to keep the orbit shape the same. The gravitational force of the comet will raise a tide on Earth much greater than lunar tides. It will affect the rotation of the earth and the orbit of the moon, but these effects can be canceled by the appropriate choices of comet orbits. Note that if our descendants have mastered the technology to alter the orbits of comets, they can also protect Earth from future impact events that could cause great extinctions. This is a project that has already received attention from government agencies.

4. How quantum effects during cosmic inflation can create density differences is explained in FAQ-10.

5. How steady expansion can go on is discussed further in chapter 8.

Chapter 6. Bringing the Universe Down to Earth

1. The origin of Scientific Mediation is described in FAQ-38. The article mentioned in the text is Nancy Ellen Abrams and R. Stephen Berry, *Bulletin of the Atomic Scientists* 33, no. 4 (1977): 50–53. For more on the Scientific Mediation in Sweden on nuclear waste and the aftermath, see Nancy E. Abrams, "Nuclear Politics in Sweden," *Environment* 21, no. 4 (1979): 6–11, 39–40. These and related articles are on the web at http://physics.ucsc.edu/~joel/abramsprimack.html.

Chapter 7. A New Origin Story

1. For more details on quantum fluctuations during cosmic inflation, see FAQ-10; on the great annihilation, see FAQ-11; on what happened in the first hundred million years, see FAQ-9. For more on how galaxies formed, see FAQ-13 and FAQ-15.

2. On how seriously to take speculations about Eternal Inflation, see FAQ-39.

Chapter 8. Cosmic Society Now

1. Martin Rees, "The Royal Society's Wider Role," *Science,* 25 June 2010, 1611.

2. For more details and references on the first six ways Earth is special, see our book *The View from the Center of the Universe: Discovering Our Extraordinary Place in the Cosmos* (New York: Riverhead, 2006), chap. 8. Hot Jupiters and massive planets more generally are much easier to find in extrasolar planetary systems than are earth-size planets, so the likelihood of earthlike planets still remains to be seen. The key paper on the Nice model is R. Gomes, H. F. Levison, K. Tsiganis, and A. Morbidelli, "Origin of the Cataclysmic Late Heavy Bombardment Period of the Terrestrial Planets," *Nature* B435 (2005): 466–469. On the larger amounts of debris and comets around most sunlike stars compared with the solar system, see J. S. Greaves and M. C. Wyatt, "Debris Discs and Comet Populations around Sun-like Stars: The Solar System in Context," *Monthly Notices of the Royal Astronomical Society* 404 (2010): 1944–1951.

3. The data on energy use per capita are from the Carbon Dioxide Information Analysis Center at Oak Ridge National Laboratory, http://cdiac.ornl.gov. The noted British economist Nicholas Stern argued that despite the urgency of limiting human-caused global warming, growth is not incompatible with sustainability. See Nicholas Stern, "Climate: What You Need to Know," *New York Review of Books,* 24 June 2010, 35–37. For more on economic growth and the environment, see FAQ-35.

4. The US population is near the bottom in accepting biological evolution: J. D. Miller, E. C. Scott, and S. Okamoto, "Public Acceptance of Evolution," *Science,* 11 August 2006, 765–766. For statements from religious authorities that biological evolution is compatible with their beliefs, see *Science, Evolution, and Creationism* (Washington, DC: National Academies Press, 2008), 12–15; and Molleen Matsumura, ed., *Voices for Evolution* (Berkeley, CA: National Center for Science Education, 1995), 84–114. See also James B. Miller, ed., *An Evolving Dialogue: Scientific, Historical, Philosophical, and Theological Perspectives on Evolution* (Washington, DC: American Association for the Advancement of Science, 1998); and Michael Ruse, *The Evolution Wars: A Guide to the Debates* (New Brunswick, NJ: Rutgers University Press, 2001).

5. That a small group of scientists with little relevant expertise but strong economic incentives has spread doubt as a political weapon is documented in Naomi Oreskes and Erik M. Conway, *Merchants of Doubt: How*

a Handful of Scientists Obscured the Truth on Issues from Tobacco Smoke to Global Warming (New York: Bloomsbury Press, 2010). See also Philip Kitcher, "The Climate Change Debates," *Science,* 4 June 2010, 1230–1234.

6. On the need and possible inspiration for a new myth, see FAQ-36.

7. On the role of metaphor, see FAQ-37. On religion, see FAQ-40.

Frequently Asked Questions

1. Clifford M. Will, *Was Einstein Right? Putting General Relativity to the Test,* 2nd ed. (New York: Basic Books, 1993). For a more up-to-date technical discussion see Clifford M. Will, "The Confrontation between General Relativity and Experiment," http://relativity.livingreviews.org/Articles/lrr-2006-3/.

2. This idea was first proposed in H. Pagels and J. R. Primack, "Supersymmetry, Cosmology, and New Physics at Teraelectronvolt Energies," *Physical Review Letters* 48 (1982): 223–226.

3. E. Komatsu et al., "Seven-Year WMAP Observations: Cosmological Interpretation," *Astrophysical Journal Supplement Series* (in press), table 2, note g.

4. Shaun A. Thomas, Filipe B. Abdalla, and Ofer Lahav, "Upper Bound of 0.28 eV on the Neutrino Masses from the Largest Photometric Redshift Survey," *Physical Review Letters* 105 (2010): 031301.

5. B. Kayser, "Neutrino Mass, Mixing, and Flavor Change," http://pdg.lbl.gov/2008/reviews/rpp2008-rev-neutrino-mixing.pdf.

6. The National Academy of Sciences National Research Council study *NASA's Beyond Einstein Program: An Architecture for Implementation* (Washington, DC: National Academies Press, 2007), on which Primack was the senior cosmologist, recommended as its highest priority such a dark energy space observatory, called the Joint Dark Energy Mission (JDEM). The study text can be downloaded from http://www.nap.edu/catalog.php?record_id=12006. Primack's invited presentation on this to the High Energy Physics Advisory Panel (HEPAP) to the Department of Energy and the National Science Foundation is at http://www.er.doe.gov/hep/files/pdfs/HEPAP-Primack.pdf. The National Academy of Sciences 2010 Decadal Survey *New Worlds, New Horizons in Astronomy and Astrophysics* (Washington, DC: National Academy Press, 2010) designated the Wide-Field Infrared Survey Telescope (WFIRST), an updated version of JDEM, as the highest-priority space mission.

7. For more on the origin of fluctuations during cosmic inflation, see Alan Guth, *The Inflationary Universe: The Quest for a New Theory of Cosmic Origins* (Reading, MA: Addison-Wesley, 1997).

8. More precise numbers are given in standard advanced textbooks, e.g., Edward W. Kolb and Michael S. Turner, *The Early Universe* (Menlo Park, CA: Addison-Wesley, 1990), chap. 6; and D. Bailin and A. Love, *Cosmology in Gauge Field Theory and String Theory* (Bristol, UK: Institute of Physics Publishing, 2004). The latter reference (on p. 92) gives the excess of matter particles over antimatter particles as 0.636×10^{-9}, with an uncertainty of about 5 percent. Modern particle physics theories satisfy the three conditions that Andrei Sakharov showed are required to explain how such a one-in-a-billion asymmetry between matter and antimatter could have arisen in the early universe.

9. *Bolshoi* is Russian for big or grand. The Bolshoi Ballet performs in the Bolshoi Theater in Moscow. Anatoly Klypin, now a professor of astronomy at New Mexico State University, was born in Moscow. For details on the Bolshoi simulation, see A. Klypin, S. Trujillo-Gomez, and J. R. Primack, "Halos and Galaxies in the Standard Cosmological Model: Results from the Bolshoi Simulation," *Astrophysical Journal,* in press (2010).

10. See James D. Barrow and Frank J. Tipler, *The Anthropic Cosmological Principle* (Oxford: Oxford University Press, 1988); and Martin J. Rees, *Just Six Numbers: The Deep Forces That Shape the Universe* (New York: Basic Books, 2000).

11. For an introduction to string theory, see Brian Greene, *The Elegant Universe: Superstrings, Hidden Dimensions, and the Quest for the Ultimate Theory* (New York: Vintage, 2000).

12. G. R. Blumenthal, S. M. Faber, J. R. Primack, and M. J. Rees, "Formation of Galaxies and Large-Scale Structure with Cold Dark Matter," *Nature* 311 (1984): 517–525.

13. See, e.g., Heinz Pagels, *The Cosmic Code: Quantum Physics as the Language of Nature* (New York: Simon and Schuster, 1982), chap. 12; and Bruce Rosenblum and Fred Kuttner, *The Quantum Enigma: Physics Encounters Consciousness* (Oxford: Oxford University Press, 2006), 186.

14. See also Gilbert Ryle, *The Concept of Mind* (New York: Hutchinson's University Library, 1949); and Antonio Damasio, *Descartes' Error: Emotion, Reason, and the Human Brain* (New York: G. P. Putnam, 1994).

15. Arthur H. Rosenfeld, "The Art of Energy Efficiency," *Annual Reviews of Energy and Environment* 24 (1999): 33–82. Also *VFTC,* 264.

16. George Lakoff and Mark Johnson, *Metaphors We Live By* (Chicago: University of Chicago Press, 1980).

Recommendations for Further Reading

Introduction to Modern Cosmology

Adams, Fred C., and Gregory Laughlin. *The Five Ages of the Universe: Inside the Physics of Eternity.* New York: Free Press, 2000. Two astrophysicists explain our best present understanding of the distant past and the distant future of the universe.

Davies, Paul. *The Last Three Minutes: Conjectures about the Ultimate Fate of the Universe.* New York: Basic Books, 1997. An entertaining discussion of ultimate cosmological questions.

Primack, Joel R., and Nancy Ellen Abrams. *The View from the Center of the Universe: Discovering Our Extraordinary Place in the Cosmos.* New York: Riverhead, 2006. An accessible presentation of modern cosmology and what it may be telling us about how we humans fit into the universe. Includes the history of earlier cosmologies, scientific details, and references.

Rees, Martin J. *Just Six Numbers: The Deep Forces That Shape the Universe.* New York: Basic Books, 2000. An introduction to the "anthropic" issues in cosmology—why the nature of the universe is determined by a small number of cosmic parameters, and why creatures like us would be impossible if any of these numbers were significantly different.

———. *Our Cosmic Habitat.* Princeton, NJ: Princeton University Press, 2001. A more informal version of the previous book based on Rees's Scribner Lectures at Princeton University.

Seife, Charles. *Alpha and Omega: The Search for the Beginning and End of the Universe.* New York: Viking, 2003. An overview of cosmology by a science journalist.

Weinberg, Steven. *The First Three Minutes: A Modern View of the Origin of the Universe.* 2nd ed. New York: Basic Books, 1993. The first popular account of modern cosmology, by a Nobel laureate who was a main creator of the Standard Model of particle physics.

Introduction to Modern Cosmology—Accessible Textbooks

Harrison, Edward R. *Cosmology: The Science of the Universe.* 2nd ed. Cambridge: Cambridge University Press, 2000. Illuminating illustrations and historical sidelights. Several possible cosmologies are discussed, but not the recent data strongly favoring the double dark theory.

Ryden, Barbara. *Introduction to Cosmology.* San Francisco: Addison-Wesley, 2003. The Double Dark theory—here called the "benchmark model"—is emphasized in this book for undergraduate physics students.

Shu, Frank. *The Physical Universe: An Introduction to Astronomy.* Sausalito, CA: University Science Books, 1982. A wonderful book for undergraduate physics students.

History of Cosmology

Ferris, Timothy. *Coming of Age in the Milky Way.* New York: Morrow, 1988. A historical introduction to the big questions of astronomy.

Goldsmith, Donald. *400 Years of the Telescope: A Journey of Science, Technology and Thought.* Chico, CA: Interstellar Studios, 2009. Excellent companion book to the PBS TV special, explaining how better telescopes have led to new discoveries about the universe.

Kuhn, Thomas S. *The Copernican Revolution: Planetary Astronomy in the Development of Western Thought.* Cambridge, MA: Harvard University Press, 1957. A classic account of the historical origins and cultural impact of the first scientific revolution.

Lemonick, Michael D. *The Light at the Edge of the Universe: Dispatches from the Front Lines of Cosmology.* Princeton, NJ: Princeton University Press, 1993. A you-are-there account of an exciting period of modern cosmology.

Overbye, Dennis. *Lonely Hearts of the Cosmos: The Story of the Scientific Quest for the Secret of the Universe.* New York: HarperPerennial, 1992. Focuses on the scientists involved in developing modern cosmology, especially Alan Sandage; includes interviews with Joel R. Primack and ends with the lyrics of a song by Nancy Ellen Abrams written for a 1986 cosmology conference.

Yulsman, Tom. *Origins: The Quest for Our Cosmic Roots.* Bristol, UK: Institute of Physics Publishing, 2003. A historical introduction by a science journalist, based in part on interviews with several of the leaders of modern cosmology.

Cosmic Inflation

Barrow, John D. *The Book of Nothing: Vacuums, Voids, and the Latest Ideas about the Origins of the Universe.* New York: Vintage Books, 2000. The history, philosophy, and science of empty space.

Davies, Paul. *Cosmic Jackpot: Why Our Universe Is Just Right for Life.* Boston: Houghton Mifflin, 2007. A whirlwind tour through competing and often wild views of the ultimate nature of our universe and other possible universes.

Ferris, Timothy. *The Whole Shebang: A State-of-the-Universe(s) Report.* New York: Simon and Schuster, 1997. An overview of modern cosmology emphasizing multiverse ideas.

Guth, Alan. *The Inflationary Universe: The Quest for a New Theory of Cosmic Origins.* Reading, MA: Addison-Wesley, 1997. An accessible introduction to cosmic inflation by its main originator.

Rees, Martin J. *Before the Beginning: Our Universe and Others.* Cambridge, MA: Helix Books, 1997. A popular introduction to cosmology and inflation theory by an eminent theoretical astrophysicist.

Vilenkin, Alex. *Many Worlds in One: The Search for Other Universes.* New York: Hill and Wang, 2006. A popular account of how our universe and others might have begun, by a theorist who originated some of these ideas.

Dark Matter

Bartusiak, Martha. *Through a Universe Darkly: A Cosmic Tale of Ancient Ethers, Dark Matter, and the Fate of the Universe.* New York: HarperCollins, 1993. A historical introduction, written before much relevant observational evidence became available.
Freeman, Ken, and Geoff McNamara. *In Search of Dark Matter.* Berlin: Springer, 2006. A historical introduction by a well-known observational astronomer and a science teacher.
Krauss, Lawrence M. *Quintessence: The Mystery of the Missing Mass.* New York: Basic Books, 2000. An introduction to dark matter and dark energy by a theoretical astrophysicist who is also a popular science writer.

Particle Physics

Feynman, Richard. *QED: The Strange Theory of Light and Matter.* Princeton, NJ: Princeton University Press, 1988. An engaging introduction to quantum electrodynamics by the Nobel laureate physicist who largely invented it.
Kane, Gordon. *The Particle Garden: Our Universe as Understood by Particle Physicists.* Cambridge, MA: Helix Books, 1995. Introduction to the Standard Model of particle physics and why it is incomplete.
———. *Supersymmetry: Unveiling the Ultimate Laws of Nature.* Cambridge, MA: Helix Books, 2000. Why supersymmetry is the best idea for going beyond the Standard Model of particle physics.
Weinberg, Steven. *Dreams of a Final Theory: The Scientist's Search for the Ultimate Laws of Nature.* New York: Pantheon Books, 1994. Historical and philosophical reflections on particle physics by a modern master.

Gravity and General Relativity

Begelman, Mitchell C., and Martin J. Rees. *Gravity's Fatal Attraction: Black Holes in the Universe.* New York: Scientific American Library, 1998. Engaging illustrated account by leading experts.
Schutz, Bernard F. *Gravity from the Ground Up: An Introductory*

Guide to Gravity and General Relativity. Cambridge: Cambridge University Press, 2003. A nonmathematical but sophisticated introduction to the many roles of gravity in the universe.

Thorne, Kip. *Black Holes and Time Warps: Einstein's Outrageous Legacy.* New York: Norton, 1994. An introduction to general relativity and modern work on it, with wonderful anecdotes.

Life in the Universe

Davies, Paul. *The Fifth Miracle: The Search for the Origin of Life.* New York: Simon and Schuster, 2000. Maybe earth life originated on Mars, and other intriguing perspectives.

Ferris, Timothy. *Life beyond Earth.* New York: Simon and Schuster, 2000. Beautifully illustrated companion volume to Ferris's PBS TV special.

Grinspoon, David. *Lonely Planets: The Natural Philosophy of Alien Life.* New York: HarperCollins, 2003. A personal view of planetary astronomy and astrobiology.

Krauss, Lawrence M. *Atom: An Odyssey from the Big Bang to Life on Earth . . . and Beyond.* Boston: Little, Brown, 2001. Follows an oxygen atom from the Big Bang to living organisms.

Lemonick, Michael D. *Other Worlds: The Search for Life in the Universe.* New York: Simon and Schuster, 1998. An overview by the former science editor of *Time.*

Morris, Simon Conway. *Life's Solution: Inevitable Humans in a Lonely Universe.* Cambridge: Cambridge University Press, 2003. A noted authority on the emergence of large organisms on earth argues that it should not surprise us that they evolved into intelligent creatures like humans.

Primack, Joel R., and Nancy Ellen Abrams. *The View from the Center of the Universe: Discovering Our Extraordinary Place in the Cosmos.* New York: Riverhead, 2006. Chapter 8 is about life in the universe.

Shostak, Seth. *Confessions of an Alien Hunter: A Scientist's Search for Extraterrestrial Intelligence.* Washington, DC: National Geographic, 2009. An engaging personal account by a senior scientist at the SETI Institute.

Ward, Peter D., and Donald Brownlee. *The Life and Death of Planet*

Earth: How the New Science of Astrobiology Charts the Ultimate Fate of Our World. New York: Henry Holt, 2002. How the increasing luminosity of the sun spells ultimate doom for the earth.

Life in the Universe—Accessible Textbooks

Goldsmith, Donald. *The Quest for Extraterrestrial Life: A Book of Readings.* Mill Valley, CA: University Science Books, 1980. A collection of classic articles, many of which remain essential.

Lunine, Jonathan Irving. *Astrobiology: A Multidisciplinary Approach.* San Francisco: Pearson Addison Wesley, 2005. From the basic science to the frontiers of research on life and planetary evolution.

Climate Change and Other Environmental Challenges

Cohen, Joel E. *How Many People Can the Earth Support?* New York: W. W. Norton, 1995. A classic summary of many approaches to the question.

Diamond, Jared. *Collapse: How Societies Choose to Fail or Succeed.* New York: Viking, 2005. Vivid accounts of a variety of environmental catastrophes, leading up to a discussion of global survival today. Diamond's answers at the end to many naysayers about our current environmental problems are especially recommended.

Gore, Al. *An Inconvenient Truth: The Planetary Emergency of Global Warming and What We Can Do about It.* Emmaus, PA: Rodale, 2006. Companion book to Gore's Oscar-winning movie, with photos and references.

———. *Our Choice: A Plan to Solve the Climate Crisis.* Emmaus, PA: Rodale, 2009. A practical illustrated guide to relevant new technologies.

Hawken, Paul. *Blessed Unrest: How the Largest Movement in the World Came into Being and Why No One Saw It Coming.* New York: Viking, 2007. Impressive evidence and analysis support this argument that there is a worldwide pattern of transformation in what seem at first to be millions of small, unrelated attempts at environmental protection or social justice.

Muller, Richard A. *Physics for Future Presidents: The Science behind the Headlines.* New York: W. W. Norton, 2008. A primer on the physics behind nuclear weapons, global warming, and other challenges, based on Muller's popular University of California, Berkeley, course.

Rees, Martin J. *Our Final Hour: A Scientist's Warning: How Terror, Error, and Environmental Disaster Threaten Humankind's Future in This Century on Earth and Beyond.* New York: Basic Books, 2003. A catalogue of the present dangers to humanity, by the eminent astrophysicist who is past president of the Royal Society.

Schneider, Stephen H. *Science as a Contact Sport: Inside the Battle to Save Earth's Climate.* Washington, DC: National Geographic, 2009. A political and scientific memoir by a leading climate scientist who died in 2010 whose work helped the Intergovernmental Panel on Climate Change share the 2007 Nobel Peace Prize with Al Gore. Describes his simultaneous struggles to figure out how climate is changing and to get the world to take action.

Speth, James Gustave. *The Bridge at the Edge of the World: Capitalism, the Environment, and Crossing from Crisis to Sustainability.* New Haven: Yale University Press, 2008. Why achieving sustainability will require major cultural and economic changes.

Stern, Nicholas. "Climate: What You Need to Know." *New York Review of Books,* June 24, 2010, 35–37. The noted British economist argues that despite the urgency of limiting human-caused global warming, growth is not incompatible with sustainability.

Taylor, Graeme. *Evolution's Edge: The Coming Collapse and Transformation of Our World.* Gabriola Island, BC: New Society Publishers, 2008. Obstacles to progress toward sustainability, and choices the world must make to overcome them.

Weart, Spencer R. *The Discovery of Global Warming: Revised and Expanded Edition.* Cambridge, MA: Harvard University Press, 2008. The history of the science of climate, by a historian of science. See also http://www.aip.org/history/climate/.

Wilson, Edward O. *The Future of Life.* New York: Vintage Books, 2002. The challenge of the twenty-first century will be "to raise the poor to a decent standard of living worldwide while preserving as much of the rest of life as possible."

Cosmology and Religion

Barbour, Ian G. *Religion and Science: Historical and Contemporary Issues.* San Francisco, CA: HarperSanFrancisco, 1997. An overview by a leading scholar, based on his Gifford Lectures.

Davies, Paul. *The Mind of God: The Scientific Basis for a Rational World.* New York: Simon and Schuster, 1993. Theological issues raised by relativity, quantum theory, and cosmology.

Ferguson, Kitty. *The Fire in the Equations: Science, Religion, and the Search for God.* New York: Bantam Press, 1994. A response to Stephen Hawking's question, in *Brief History of Time,* "What is it that breathes fire into the equations and makes a universe for them to describe?"

Frank, Adam. *The Constant Fire: Beyond the Science vs. Religion Debate.* Berkeley: University of California Press, 2009. Cosmology as sacred narrative, by a literary astrophysicist.

Jammer, Max. *Einstein and Religion: Physics and Theology.* Princeton, NJ: Princeton University Press, 1999.

Matt, Daniel C. *God and the Big Bang: Discovering Harmony between Science and Spirituality.* Woodstock, VT: Jewish Lights, 1996. A Jewish mystical perspective on modern cosmology.

Wilson, David Sloan. *Darwin's Cathedral: Evolution, Religion, and the Nature of Society.* Chicago: University of Chicago Press, 2002. An evolutionary account of religion emphasizing how it promotes social cohesiveness.

Developing a New Unifying Myth

Berry, Thomas. *The Dream of the Earth.* 2nd ed. San Francisco: Sierra Club Books, 2006. A poetic and prophetic book about the origin of our global environmental challenges and the cultural and spiritual changes required to address them.

Campbell, Joseph. *The Inner Reaches of Outer Space: Metaphor as Myth and as Religion.* New York: A. van der Marck Editions, 1986. The last book by the famous mythologist seeks a new globally shared origin story based on modern science.

Kauffman, Stuart A. *Reinventing the Sacred: A New View of Science,*

Reason, and Religion. New York: Basic Books, 2008. Suggesting that the natural creativity of the universe, which reveals itself through the emergence of increasing complexity, is a new kind of sacredness.

Wright, Robert. *The Evolution of God*. New York: Little, Brown, 2009. How religion has embraced increasingly inclusive groups, and how further inclusivity can save us all.

———. *Nonzero: The Logic of Human Destiny*. New York: Vintage, 2001. Why natural selection and human cultural evolution favor win-win situations, and how cooperation can help us reach sustainability.

About the Illustrations

▣ This symbol indicates that a related video can be watched at http://new-universe.org, where additional videos, information about the illustrations, and links to their sources can also be found.

Chapter 1. The New Universe

Figure 1. *Great telescopes on and orbiting Earth.* Image: Nina McCurdy.

Figure 2. *A pillar of star birth: The Carina Nebula in visible light.* This image shows the tip of the three-light-year-long pillar bathed in the glow of light from hot, massive stars off the top of the image. Scorching radiation and fast winds (streams of charged particles) from these stars are sculpting the pillar and causing new stars to form within it. Streamers of gas and dust can be seen flowing off the top of the structure. Dust hides much of the nebula. This photo was taken by Wide Field Camera 3 on the Hubble Space Telescope. Image and description: NASA, ESA, and the Hubble SM4 ERO Team.

Figure 3. *A pillar of star birth: The Carina Nebula in infrared light.* In this photo, also taken by the Wide Field Camera 3 (WFC3), infrared light penetrates the dust, making it possible to see far more stars. The WFC3, installed by astronauts during the final Hubble Space Telescope service mission in May 2009, tremendously increased the telescope's infrared capability. Image: NASA, ESA, and the Hubble SM4 ERO Team.

Figure 4. *Saturn with Earth in the background.* This photo is from the Cassini spacecraft orbiting Saturn. Image: NASA/JPL/Space Science Institute.

Figure 5. *The Hubble Ultra Deep Field in infrared light.* This photo, taken by the Wide Field Camera 3 on the Hubble Space Telescope, is discussed further at the beginning of chapter 4. Image: NASA, ESA, G. Illingworth, and R. Bouwens (University of California, Santa Cruz), and the HUDF09 Team.

Figure 6. *The ancient Egyptian cosmos.* Vignette from the Book of the Dead of Nesitanebtashru (a 21st Dynasty priestess), Egypt, 1025 BC. Image: © 2010 The British Museum Images. All rights reserved. Used by permission.

Figure 7. *The ancient Egyptian cosmos, simplified version.* Image: Nicolle Rager Fuller.

Figure 8. *The ancient Hebrew cosmos.* Image: Nina McCurdy.

Figure 9. *The medieval cosmos.* We have drawn the spheres here as the people of the Middle Ages described them—and as they would have drawn them, had they known how to draw in perspective. Image: Nicolle Rager Fuller.

Figure 10. *The Newtonian cosmos, as represented by M. C. Escher's "Cubic Space Division."* M. C. Escher's *Cubic Space Division* © 2010 The M. C. Escher Company—Holland. All rights reserved. www.mcescher.com. Used by permission.

Figure 11. *William Herschel's map of the Milky Way.*

Figure 12. *Our address in the universe.* Galaxy image courtesy NASA Images; Sloan Digital Sky Survey (SDSS) data courtesy SDSS. Image: Nicolle Rager Fuller/Nina McCurdy/NASA/JPL-Caltech/M. Tegmark & the SDSS Collaboration, www.sdss.org.

▣ *Voyage to the Virgo Cluster* video at http://new-universe.org.

Figure 13. *The Orion Constellation as seen from Earth.* From the *Voyage to the Virgo Cluster* video. Image: Courtesy PBS NOVA television/Donna Cox/Stuart Levy.

Figure 14. *The Orion Nebula.* This image was taken by the Advanced Camera for Surveys on the Hubble Space Telescope. Image: NASA, ESA, M. Robberto (Space Telescope Science Institute/ESA), and the Hubble Space Telescope Orion Treasury Project Team.

Figure 15. *The Milky Way Galaxy with large and small Magellanic clouds.* Collage by Nina McCurdy, including Nick Risinger's *Artist's Conception of the Milky Way Galaxy,* adapted from NASA images.

Figure 16. *The Virgo Cluster and a chain of galaxies.* Image from the *Voyage to the Virgo Cluster* video. The chain galaxies are mostly in the Ursa Major Groups. Image: Courtesy PBS NOVA television/Donna Cox/Stuart Levy.

Figure 17. *The Whirlpool Galaxy (M51).* This photo was taken by the Hubble Space Telescope. Image: NASA, ESA, S. Beckwith, and the Hubble Heritage Team (STScI/AURA).

Figure 18. *Galaxy M87 at the center of the Virgo Cluster.* This photo was taken by the Hubble Space Telescope. Image: NASA, ESA, and the Hubble Heritage Team (STScI/AURA).

Figure 19. *Peanuts: "You are of no significance."* Peanuts © 1997, United Features Syndicate, Inc. Used by permission.

Figure 20. *Calvin and Hobbes: "What a clear night!"* CALVIN AND HOBBES © 1988 Watterson. Dist. by UNIVERSAL UCLICK. Reprinted with permission. All rights reserved.

Chapter 2. Size Is Destiny

Figure 21. *Sizes are doorways within doorways.* Image: Nicolle Rager Fuller.

Figure 22. *The Cosmic Uroboros.* The uroboros ranges from the smallest size, the Planck size (10^{-33} cm) at the tip of the tail, to the size of the entire visible universe (10^{29} cm) at the head of the serpent. In modern physics, forces result from the exchange of particles. The photon, the particle of light, is responsible for electrical and magnetic forces. Analogous particles called gluons carry the strong force, which holds protons, neutrons, and the entire nucleus together. The weak interactions, which are responsible for certain kinds of radioactive decays, are due to the exchange of massive W and Z particles. In a Grand Unified Theory (GUT), all these forces come together and have the same strength on very small scales, represented by "GUT" on the serpent's tail. Image: Nicolle Rager Fuller.

▣ *Powers of ten zoom,* from *Cosmic Voyage* IMAX film, video at http://new-universe.org.

Chapter 3. We Are Stardust

Figure 23. *The Pyramid of All Visible Matter.* Image: Nicolle Rager Fuller.

Figure 24. *The periodic table of the elements, with the origins of each element.* Image: Nina McCurdy.

Figure 25. *Kepler's supernova remnant, from the explosion of a white dwarf.* This is a composite image from the Chandra X-ray and Spitzer Infrared Space Telescopes. Image: NASA/ESA/JHU/R. Sankrit & W. Blair.

Figure 26. *The Crab Nebula, remnant of the explosion of a massive star.* This image was taken by Wide Field Planetary Camera 2 on the Hubble Space Telescope. X-ray: NASA/CXC/ASU/J. Hester et al.; Optical: NASA/ESA/ASU/J. Hester & A. Loll; Infrared: NASA/JPL-Caltech/Univ. Minn./R. Gehrz.

▣ *Zooming in to the Cat's Eye Nebula* video at http://new-universe.org.

Figure 27. *A large-scale view of the Cat's Eye Nebula.* Image: Nordic Optical Telescope and Romano Corradi (Isaac Newton Group of Telescopes, Spain).

Figure 28. *The Cat's Eye Nebula.* This image was taken by the Advanced Camera for Surveys on the Hubble Space Telescope. Image: NASA, ESA, HEIC, and the Hubble Heritage Team (STScI/AURA).

Figure 29. *Manorbier Beach, Pembrokeshire, Wales.* John Williamson Photography. Used by permission.

Figure 30. *The Cosmic Density Pyramid.* Image: Nicolle Rager Fuller.

Figure 31. *Dark matter ships on an ocean of dark energy.* Image: Garth von Ahnen.

Figure 32. *Big Bang data agree with the Double Dark theory.* The blue curve is the prediction of the Double Dark theory, which agrees spectacularly well with the white points, the 2010 data from NASA's Wilkinson Microwave Anisotropy Probe satellite. What is graphed is the amount of structure on different angular scales in the temperature of the cosmic background radiation, pictured on the sphere. Large angular scales are on the left, and smaller ones are on the right. The graph is from N. Jarosik et al., "Seven-Year Wilkinson Microwave Anisotropy Probe (WMAP7) Observations: Sky Maps, Systematic

Errors, and Basic Results, *Astrophysical Journal Supplement* (in press). Ground-based data are from the ACBAR and QUaD experiments at the South Pole, pictured in the figure. The Cosmic Background Explorer (COBE) satellite provided data on the largest angular scales, and COBE data announced in 1992 gave an early confirmation of a key Cold Dark Matter theory prediction.

Figure 33. *The distribution of matter also agrees with the Double Dark theory.* The same theory that correctly predicts the properties of the cosmic background radiation (the blue curve in figure 32) also correctly predicts the amount of structure on different length scales in the universe today (red curve). Plot by Max Tegmark, MIT. Courtesy M. Tegmark and the SDSS Collaboration, www.sdss.org. Used by permission.

Figure 34. *"Don't Feel Bad, Loretta . . . The Entire Universe Is Expanding."* Lockhorns © 1995 Wm. Hoest Enterprises, Inc. King Features Syndicate. Used by permission.

Figure 35. Still image from ▣ *A simulation of the expansion of the universe.* Simulation by Ben Moore. Used by permission.

Figure 36. *The end of expansion.* Illustrated with simulation by Ben Moore. Used by permission.

Figure 37. *Wild space, tame space.* Illustrated with simulation by Ben Moore. Used by permission.

Figure 38. *The evolution of the cosmic web.* Supercomputer simulations carried out by Anatoly Klypin and Andrey Kravtsov at the National Center for Supercomputing Applications (NCSA). Used by permission.

▣ *Columbia Simulation* video at http://new-universe.org.

▣ *Bolshoi Zoom-In* video at http://new-universe.org.

▣ *Bolshoi Fly-Through* video at http://new-universe.org.

Figure 39. *The Bolshoi simulation—one billion light-years across.* This image shows the distribution of dark matter according to a recent supercomputer simulation. For an explanation of why the dark matter distribution looks so filamentary, see FAQ-12. Image: Anatoly Klypin, Stefan Gottlöber, and Joel R. Primack.

▣ *Aquarius Simulation* video at http://new-universe.org.

Figure 40. *The Aquarius simulation of a Milky Way–size dark matter halo.* Courtesy of Volker Springel, Max Planck Institute for Astrophysics, Garching, Germany. Used by permission.

Figure 41. *The Eye of the Pyramid of All Visible Matter.* Detail from figure 23.

Chapter 4. Our Place in Time

Figure 42. *The Hubble Ultra Deep Field in optical light.* This photo was taken by the Advanced Camera for Surveys on the Hubble Space Telescope. The Ultra Deep Field observations represent a narrow, deep view of the cosmos. Peering into the Ultra Deep Field is like looking through an eight-foot-long soda straw. In ground-based photographs, the patch of sky in which the galaxies reside (just one-tenth the diameter of the full moon) is so empty that only a handful of stars (with four-pointed spikes in the image) within the Milky Way galaxy can be seen. Image: NASA, ESA, S. Beckwith (STScI), and the HUDF Team.

Figures 43, 44, and 45. *Zooming in to the Hubble Ultra Deep Field.* Snapshots taken from ▣ *Hubble Ultra Deep Field Zoom-In* video. Figure 43 is a view of galaxies whose light has been on its way to us at least as long as Earth has existed. In the center right of figure 43 there is an elliptical galaxy in front of a spiral galaxy. Figure 44 looks back to the first three billion years, and figure 45 looks back to the first billion years. Images: NASA, ESA, F. Summers, Z. Levay, L. Frattare, B. Mobasher, A. Koekemoer, and the HUDF Team (STScI).

Figure 46. *The Sloan Digital Sky Survey.* This still image is from the ▣ *Galaxies Mapped by the Sloan Digital Sky Survey* video. Courtesy Mark SubbaRao and Dinoj Surendran, Adler Planetarium/University of Chicago. Used by permission.

Figure 47. *The cosmic microwave background radiation sphere.* This still image is from the ▣ *Galaxies Mapped by the Sloan Digital Sky Survey* video. Courtesy Mark SubbaRao and Dinoj Surendran, Adler Planetarium/University of Chicago. Used by permission.

Figure 48. *The Cosmic Spheres of Time.* Image: Nicolle Rager Fuller.

Chapter 5. This Cosmically Pivotal Moment

Figure 49. *Two galaxies merging.* These four snapshots are from ▣ a high-resolution hydrodynamical simulation done by Patrik Jonsson, Greg Novak, and Joel Primack using Volker Springel's simulation code GADGET and Patrik Jonsson's *Sunrise* code to visualize the effects of stellar evolution, dust scattering and absorption of light, and re-radiation of the energy at longer wavelengths. *Upper left:* galaxies first approach; *upper right:* they then separate; *lower left:* gravity brings them back together and their centers merge; *lower right:* an elliptical galaxy results. Collage created by Nina McCurdy.

Figure 50. *The changing luminosity of the sun.* Image: Nina McCurdy, using data from I.-Juliana Sackmann, Arnold I. Boothroyd, and Kathleen E. Kraemer, "Our Sun. III. Present and Future," *Astrophysical Journal* 418 (1993): 457.

Figure 51. *Human population growth.* Image: Nicolle Rager Fuller/Nina McCurdy.

Figure 52. *The exponential growth of pond scum.* Image: Nina McCurdy.

Figure 53. *The daily consumption of resources per person in the United States.* Image: Nicolle Rager Fuller, using data from the 2005 Statistical Abstract of the United States.

Figure 54. *World emissions of greenhouse gases per capita in 2005.* The color code represents tons of carbon dioxide equivalent per capita per year. Image: Emission Data for Global Atmospheric Research (EDGAR).

Figure 55. *Cosmic inflation and cosmic expansion.* Image: Nicolle Rager Fuller.

Figure 56. *"Now Playing—A Reassuring Lie."* Copyright © Clay Bennett. Used by permission.

Chapter 6. Bringing the Universe Down to Earth

Figure 57. *The concentration of carbon dioxide in the atmosphere, with the exponentially growing human contribution.* Image: Nina McCurdy, using data from US Global Change Research Program (www.globalchange.gov).

Figure 58. *Projected carbon emissions through 2100, and actual data so far.* The pessimistic red curve represents business as usual (IPCC 2007 scenario A2), and the optimistic blue curve represents an aggressive reduction in carbon emissions (IPCC 2007 scenario B1). From US Global Change Research Program (www.globalchange.gov) with data so far from Carbon Dioxide Information Analysis Center, Oak Ridge National Laboratory (cdiac.ornl.gov).

Figure 59. *Optimistic and pessimistic temperature scenarios.* These scenarios correspond to the two scenarios presented in figure 58. Images are from the US Global Change Research Program (www.globalchange.gov).

Figure 60. *Space debris in Low Earth Orbit.* Low Earth Orbit, space within 2,000 kilometers of Earth's surface, is where space debris is most concentrated. Only about 5 percent of the objects in this illustration are functional satellites. NASA illustration courtesy NASA Orbital Debris Program Office.

Chapter 7. A New Origin Story

Figure 61. John Andrews, *Boston Tea Party—Destruction of the Tea in Boston Harbor, December 16, 1773.* Courtesy Yale University Art Gallery, Gift of Mr. and Mrs. Melville Chapin.

Figure 62. John Trumbull, *The Declaration of Independence, July 4, 1776.* Courtesy Yale University Art Gallery, Trumbull Collection.

Figure 63. *Cosmic microwave background radiation.* Image: NASA.

Figure 64. *A forming star.* Simulation: Tom Abel (KIPAC, Stanford University), Greg Bryan (Columbia University), and Michael Norman (University of California, San Diego, SDSC). Visualization: Ralf Kaeler and Tom Abel (KIPAC). Used by permission.

Figure 65. *A forming planet.* "This artist's conception shows the closest known planetary system to our own, called Epsilon Eridani. Observations from NASA's Spitzer Space Telescope show that the system hosts two asteroid belts, in addition to previously identified candidate planets and an outer comet ring." Images and description: NASA/JPL-Caltech.

Figure 66. *A pensive gorilla.* © iStockphoto.com/Richard Stern (rickyste). Used by permission.

Figure 67. Auguste Rodin's *The Thinker.* Photo: Nina McCurdy.

Figure 68. *Cosmic Las Vegas.* Image: Nicolle Rager Fuller.

Figure 69. *Many universes in eternal inflation.* Still from *Eternal Inflation Visualization* video. Image: Nina McCurdy/Anthony Aguirre/Nancy Abrams/Joel Primack.

▣ *Eternal Inflation Visualization* video at http://new-universe.org.

Chapter 8. Cosmic Society Now

Figure 70. *The human identity uroboros.* Image: Nina McCurdy.

Figure 71. *Child and cosmos.* Photo of Earth: Copyright © Planetary Visions Ltd., with thanks to Kevin M. Tildsley. Used by permission. Photo of child: Nancy Ellen Abrams. Collage: Nina McCurdy.

▣ *From Eternal Inflation to Earth* video at http://new-universe.org.

Frequently Asked Questions

Figure 72. *The merging history of a large dark matter halo.* Time is increasing to the right. Radii of dark matter halos are represented by the blue circles, and the size of their core regions is represented by the red dots. Many little halos at early times merge to form larger halos, which subsequently merge into one big halo. Image: Risa H. Wechsler, James S. Bullock, Joel R. Primack, Andrey V. Kravtsov, and Avishai Dekel, "Concentrations of Dark Halos from Their Assembly Histories," *Astrophysical Journal* 566 (2002): 52–70, fig. 2.

Index

Page numbers in boldface type indicate illustrations.